Handbook of Hidden Markov Models in Bioinformatics

CHAPMAN & HALL/CRC
Mathematical and Computational Biology Series

Aims and scope:

This series aims to capture new developments and summarize what is known over the whole spectrum of mathematical and computational biology and medicine. It seeks to encourage the integration of mathematical, statistical and computational methods into biology by publishing a broad range of textbooks, reference works and handbooks. The titles included in the series are meant to appeal to students, researchers and professionals in the mathematical, statistical and computational sciences, fundamental biology and bioengineering, as well as interdisciplinary researchers involved in the field. The inclusion of concrete examples and applications, and programming techniques and examples, is highly encouraged.

Series Editors

Alison M. Etheridge
Department of Statistics
University of Oxford

Louis J. Gross
Department of Ecology and Evolutionary Biology
University of Tennessee

Suzanne Lenhart
Department of Mathematics
University of Tennessee

Philip K. Maini
Mathematical Institute
University of Oxford

Shoba Ranganathan
Research Institute of Biotechnology
Macquarie University

Hershel M. Safer
Weizmann Institute of Science
Bioinformatics & Bio Computing

Eberhard O. Voit
The Wallace H. Couter Department of Biomedical Engineering
Georgia Tech and Emory University

Proposals for the series should be submitted to one of the series editors above or directly to:
CRC Press, Taylor & Francis Group
Albert House, 4th floor
1-4 Singer Street
London EC2A 4BQ
UK

Published Titles

Bioinformatics: A Practical Approach
Shui Qing Ye

Cancer Modelling and Simulation
Luigi Preziosi

Computational Biology: A Statistical Mechanics Perspective
Ralf Blossey

Computational Neuroscience: A Comprehensive Approach
Jianfeng Feng

Data Analysis Tools for DNA Microarrays
Sorin Draghici

Differential Equations and Mathematical Biology
D.S. Jones and B.D. Sleeman

Exactly Solvable Models of Biological Invasion
Sergei V. Petrovskii and Bai-Lian Li

Handbook of Hidden Markov Models in Bioinformatics
Martin Gollery

Introduction to Bioinformatics
Anna Tramontano

An Introduction to Systems Biology: Design Principles of Biological Circuits
Uri Alon

Knowledge Discovery in Proteomics
Igor Jurisica and Dennis Wigle

Modeling and Simulation of Capsules and Biological Cells
C. Pozrikidis

Niche Modeling: Predictions from Statistical Distributions
David Stockwell

Normal Mode Analysis: Theory and Applications to Biological and Chemical Systems
Qiang Cui and Ivet Bahar

Pattern Discovery in Bioinformatics: Theory & Algorithms
Laxmi Parida

Spatiotemporal Patterns in Ecology and Epidemiology: Theory, Models, and Simulation
Horst Malchow, Sergei V. Petrovskii, and Ezio Venturino

Stochastic Modelling for Systems Biology
Darren J. Wilkinson

The Ten Most Wanted Solutions in Protein Bioinformatics

Chapman & Hall/CRC Mathematical and Computational Biology Series

Handbook of Hidden Markov Models in Bioinformatics

Martin Gollery

CRC Press
Taylor & Francis Group
Boca Raton London New York

CRC Press is an imprint of the
Taylor & Francis Group, an **informa** business

A CHAPMAN & HALL BOOK

CRC Press
Taylor & Francis Group
6000 Broken Sound Parkway NW, Suite 300
Boca Raton, FL 33487-2742

First issued in paperback 2019

© 2008 by Taylor & Francis Group, LLC
CRC Press is an imprint of Taylor & Francis Group, an Informa business

No claim to original U.S. Government works

ISBN-13: 978-1-58488-684-6 (hbk)
ISBN-13: 978-0-367-38719-8 (pbk)

Library of Congress Cataloging-in-Publication Data

Gollery, Martin.
   Handbook of hidden Markov models in bioinformatics / author, Martin Gollery.
      p. cm. -- (Chapman & Hall/CRC mathematical and computational biology series)
      Includes bibliographical references and index.
   ISBN 978-1-58488-684-6 (hardback : alk. paper) 1. Bioinformatics. 2. Computational biology. 3. Markov processes. I. Title. II. Series.

QH324.2.G65 2008
570.285--dc22
                                       2008012303

Visit the Taylor & Francis Web site at
http://www.taylorandfrancis.com

and the CRC Press Web site at
http://www.crcpress.com

Dedication

No project of this size can be done alone, especially when there are children and pets at home. Fortunately, I have a partner who is an award-winning biology instructor and a brilliant scientist as well as a delightful person. My wife and best friend is Suzanne Gollery, who takes care of everything and provides unwavering support, whether it be sharing presentation duties in a bioinformatics tutorial, or sharing responsibilities when getting the children ready for school. It is to her that this book is dedicated.

Contents

Foreword

In the past ten years I have encountered numerous students and researchers who have asked me for instruction on the basics of profile Hidden Markov Models and their use in bioinformatics. The lectures that were presented in this area were well suited to those of a more mathematical bent, but baffling to those who actually wanted to use HMMs to analyze data. As a result, the use of profile HMMs has been slowed somewhat, as people tend to hesitate when asked to use what they do not understand. Eventually, I condensed this lecture into a formal presentation at the request of Dr. Jeff Harper to present to a group of his graduate students. This grew to become a tutorial at the ISMB conference, and I was then asked to expand it into a book. It is my hope that you find it useful in your future work.

Preface

This book is designed for intermediate to advanced bioinformatics students who have taken a couple of introductory courses, along with the prerequisite biochemistry and computer science introductions. I have included problem sets with each chapter with the goal of spurring further study and inquiry.

This book is also designed to be an aid to those researchers who want to get more from their data than may be seen with a standard BLAST search, and wish to obtain the greatest possible information in the least amount of time. Many useful resources in terms of databases and programs are woefully underutilized by the bioinformatics community, simply due to a lack of knowledge. Prudent use of these tools can be a tremendous benefit to most projects, whether the project is a metagenomics study or a search for function in a set of newly sequenced ESTs.

Acknowledgments

I would like to thank my collaborators and mentors at the universities and institutes around the world, from whom I have learned so much. Particular thanks to Dr. Brian Beck, who was my partner in the creation of the Nevada Center for Bioinformatics, and who is now at Texas Woman's University. I thank all my friends at the former McDonnell Douglas Astronautics company (now Boeing Space Systems), and Chevron, where I learned how to work with computers in a scientific environment. Many kudos go out to Lisa Felske and the faculty of HSPVA, and the hundreds of students who I have taught over the years. I would like to thank Richard Offerdahl, for taking a chance and hiring me into my first bioinformatics position many years ago. Kudos to Dave Rector for working with me on countless interesting problems over the years.

A special thanks goes out to the Saturday Night Alive Folk Mass Band, Donna Axton and the Toccata group for providing me with musical distractions when I needed them.

Most of all, I would like to thank my family: my parents for always acting like I could do whatever I wanted to do; my children, Sharon, Mariposa and Daniel for being the best kids in the whole wide world; and especially my wife and best friend Suzanne, who takes care of everything and provides unwavering support. It is to her that this book is dedicated.

About the Author

Martin Gollery started his research career synthesizing and analyzing electro-conducting polymers. He moved into scientific computing in the early 1980's and never quite left the field. Marty taught at the high school and college levels for a number of years, and won several awards for science education. Leaving the academic world, he joined TimeLogic, a small bioinformatics company, at a time when this was a nascent but rapidly growing field. It was here that he wrote the first papers and posters on TeraBlast, as well as beginning his work on HMMs. He moved to the University of Nevada to do research on a wide variety of organisms, and has worked as an editor and reviewer on several journals. Returning to industry, he is now a Senior Bioinformatics Scientist with Active Motif, Inc.

Along the way Marty has sung with Jazz bands, musical theater, and classical groups. He has repaired houses for the poor and built outhouses on remote reservations. He has hiked 2760 miles along the Pacific Crest Trail from Mexico to Canada, and was on the first ascents of mountains such as the Jade Dragon Mountain and Dzambala III in China.

Chapter 1

Introduction to HMMs and Related Profile Methods

1.1 INTRODUCTION

Chapter 1 provides an introduction to HMMs and related Profile methods. This chapter presents the problem of standard pairwise algorithms such as BLAST or FASTA, and then demonstrates ways in which the various profile methods provide a better solution. This chapter provides the basic understanding of what a Hidden Markov Model really is, in terms of bioinformatics, and why they are used instead of other methods. We will cover some of the background about how HMMs came to be used for biological sequence analysis, but will not spend too much time on the underlying mathematical theory. For those who are interested in the underlying basis of Hidden Markov Models and their use in bioinformatics, I recommend the other excellent books on the subject, particularly "Biological Sequence Analysis: Probabilistic Models of Proteins and Nucleic Acids" by Durbin, Eddy, Krogh and Mitchison, and "Hidden Markov Models of bioinformatics" by Koski. Particularly for the more mathematically inclined or for those interested in developing new algorithms, these books will provide an understanding of the equations and calculations that will not be covered here.

This book is for the student or working researcher who needs to learn what programs are available, where they can get them, how they can use them and why they might want to use one instead of the other. Of these topics, the emphasis will always be on how to use the programs. While many people are now using HMMs for a variety of purposes, the optimal results are frequently missed simply because the best options are not chosen. Sometimes even the program or the database is not the best fit for the problem at hand because the program is not well understood. These problems will be addressed, as well as some of the more prevalent questions that tend to crop up.

Perhaps the most common question is, 'why are they called *Hidden* Markov Models'? Let us examine the case of the muddy footprints to find out. My son Daniel walked in the door with mud on his shoes. I might assume that it is probably raining outside, but this is not necessarily the case. After all, it may have stopped raining outside, and the weather could be cloudy or even sunny,

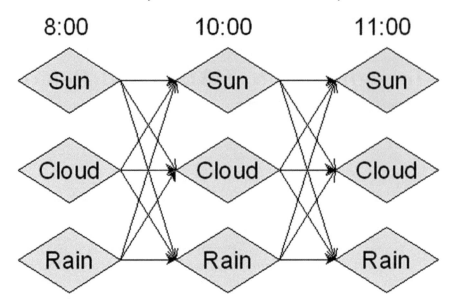

FIGURE 1.1:   Weather States

with puddles still left around. If we assume that these three conditions—sunny, cloudy or rainy—are the only possible conditions, then the probability that it is sunny, which we will call Ps, plus the probability that it is cloudy (Pc), plus the probability that it is rainy (Pr) must equal one. These values are not necessarily equal, however. Chances are that it is rainy outside, so Pr is likely greater than Ps. However, we cannot tell for certain from the observation of the mud what the weather is outside. We might say that the muddy shoes are the observation, or the *emission*, from the model. What state generated that emission is hidden from us.

If Daniel comes in the house three times today, with progressively soggier shoes each time, at 8:00, 9:00 and 10:00, I might begin to get angry, but I could represent the situation with a Hidden Markov Model with all three states possible at all three time points, for a total of 9 states and $3^3 = 27$ different possible paths through the model. One example of these paths might be rain at 8:00, clouds at 10:00 and rain again at 11:00. Another might be clouds at 8:00, sun at 10:00 and rain at 11:00.

Since I am locked up indoors with this computer, the only observations that I can make are the conditions of Daniel's shoes at each of these time points. The weather conditions are unknown to me; hence this is a Hidden Markov Model. The condition of Daniel's shoes, whether they are muddy, damp or dry, is called the emission state at that time point. Some state in the model produced the emission state of the shoes, although we do not know for certain what the state of the model was.

Okay, but who was Markov? Andrei Markov was a Russian mathematician who studied probability theory about a hundred years ago. He developed the theory of the Markov chain, which involves variables that depend on the present state, but is independent of the way that the present state evolved from its predecessors.

In the case of our simple model, we might say that the Markov chain of the weather indicates that the state at 11:00 depends on the weather at 10:00, but not on the conditions at 8:00. If it is raining at 10:00, it is likely to still be raining at 11:00 no matter what the conditions were at 8:00.

Another illustration of Markov chain theory is frequently given as a baseball game. The possible states in this analogy are the number of runners on base and the bases that they occupy, and the number of outs. If a batter is up and there are runners on first and third with two outs, then the batter is concerned only with these facts. There are indeed many different ways that this situation could have come about, but how it came to be is unimportant at this time. The only question for the batter is how to maximize the probability of increasing the score without getting a third 'out'!

1.2   INTRODUCTION TO SEQUENCE ANALYSIS

This is all well and good for muddy shoes and baseball, but what about predicting protein families? To see how this relates, we need to step back and take a look at sequence analysis at a simpler level.

Homology searching begins with the presupposition that if two genes have similar sequences, then they are likely to have similar functions. This assumption, while not always correct, can save a good deal of work, as it is much easier to compare sequences than it is to characterize the gene in the wet lab. While we will not explore the pairwise comparison methods in depth, we will cover the basic concepts in order to understand their benefits and limitations, and to see how some of these concepts apply to Profile-HMM methods.

How did we get here in the first place? That is to say, why are these sequences not identical if the genes have the same purpose? The answer lies in the evolution of the organism. If some ancestral organism has a gene that makes a protein with the sequence

MEEPQSDPSVEPPLSQETFSDLWKLLPENNVLSPLP

then that gene will be copied every time the cell undergoes mitosis, the process of duplication. The gene will be copied perfectly nearly all of the time. On very rare occasions, an error will be made in the copy, and the resultant protein can change. The daughter cell will have a protein that has somewhat different properties from the parent. The amount of change can vary dramatically

Human
MEEPQSDPSVEPPLSQETFSDLWKLLPENNVLSPLP
Chimpanzee
MEEPQSDPSVEPPLSQETFSDLWKLLPENNVLSPLP
Rat
MEDSQSDMSIELPLSQETFSCLWKLLPPDDIL
Mouse
MEESQSDISLELPLSQETFSGLWKLLPPEDIL

FIGURE 1.2:   Similar Sequences from Four Organisms

depending on the type and location of the amino acid that is altered. Some changes may be fatal to the organism, while others may be highly beneficial.

Over a long period of time, the organism may diverge into subspecies. Changes in each of these subspecies are isolated from one another, and the sequences of the genes will become increasingly divergent as time progresses. The proteins will likely diverge in function as well over evolutionary time, and the time since the split between the organisms may be estimated from the amount of diversity. Sequences of proteins that are shared between humans and chimps, for example, are typically very similar as seen in figure 1.2.

Paralogs are homologous genes within an organism that have diverged through gene duplication, as seen in figure 1.3. Orthologs are homologs in different species that derived from a common ancestor.

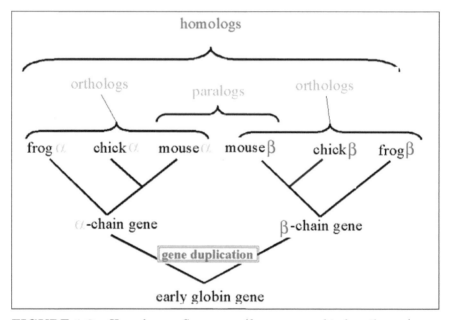

FIGURE 1.3:   Homologous Sequences (from www.ncbi.nlm.nih.gov/Education)

```
Q   1 MEEPQSDPSVEPPLSQETFSDLWKLLPENNVLSPLP-SQAMDDLMLSPDDIEQWFTEDPG
      |||||||| |·| |||||||||||||||| ||||| || | ....| || ...  |  · |
T   1 MEEPQSDLSIELPLSQETFSDLWKLLPPNNVLSTLPSSDSIEELFLS-ENVTGWLEDSGG
```

FIGURE 1.4: A Global Alignment

```
ATTGTCGCGCTTTAGAGA
||||||||||||
GCCGTCGCGCTTTA---G
```

FIGURE 1.5: A Poor Global Alignment

1.3 PAIRWISE ALGORITHMS—SMITH-WATERMAN, FASTA and BLAST

In the beginning, there was the Needleman-Wunsch algorithm. This algorithm used a technique called dynamic programming to align one gene sequence to another across the entire length of both genes. This provided a global alignment, one that aligned the full sequence of one gene or protein to the full sequence of another. To accomplish this, Needleman-Wunsch forces the end of one sequence to match the end of the other, as seen in figure 1.4.

This works very well for closely related sequences, and was certainly faster and easier than aligning by hand, but there were many times when one of the sequences was only a partial gene, or for some other reason did not align well globally to the other. For example, in the following pairwise alignment (figure 1.5), we can see that the three characters on the left do not match at all, and the G on the right would match nicely if the gaps were left out.

In the early 1980's, Temple Smith and Michael Waterman made a slight, but very important improvement to the Needleman-Wunsch algorithm by introducing a term that allowed for optimal local alignments, as in the nucleotide alignment shown in figure 1.6.

The Smith-Waterman algorithm is capable of producing much more useful alignments than Needleman-Wunsch, but could be quite slow if one needed to search an entire database of sequences. As the databases began to grow (and grow, and grow), this problem became more and more apparent. To

FIGURE 1.6: A Local Alignment Provides a Better Result

speed things up a bit, Bill Pearson introduced a heuristic approximation to the Smith-Waterman algorithm that also produced local alignments but was considerably faster. Pearson's algorithm was, appropriately enough, named FASTA due to the extra speed provided by these heuristics. As the databases continued to grow, however, the need for even more rapid methods led Warren Gish, David Lipman and Stephen Altschul to join up with Miller from Penn State and Gene Myers from the University of Arizona to develop the Basic Local Alignment Search Tool, or BLAST. This has become the most popular alignment tool in bioinformatics, and the BLAST paper is the most cited paper in any field of science over the past ten years.

BLAST relied on work by Karlin and Altschul to assess the statistical significance of the alignment. This made it easier to understand whether or not a hit was likely to be found by random chance. Without a method to assess significance, the quality of the hit is suspect, because if you search a big enough database, there will be matches to something with any sequence that is chosen, just by random chance. These statistics are shown as expectation values, also known as E-values. E-values show the expectation that a match was found by random chance. An E-value of 10 means that there would likely be 10 hits of this score simply by random chance in a database of this size, which means that there is a good chance that this hit is a false positive. On the other hand, an E-value of 1e-6 means 1×10^{-6} or, in other words, there is a one in a million chance that this hit was found by random chance. BLAST and other programs will sometimes return an E-value of 0. This is not strictly accurate because there is always some chance that this hit was found randomly, but the program returns a zero beyond a certain point, indicating that the odds that this is a random hit are very, very low.

Development on BLAST has continued with gapped alignments added in version 2.0 (shown in figure 1.7), increased speed through the use of larger seed sizes with MegaBlast in the nucleotide searches, and multiple hits in the protein searches. Other groups contributed distributed versions such as MPI-BLAST, and hardware accelerated versions such as RC-BLAST and Tera-BLAST^{TM}. A program called PatternHunter improved on the formulation of the intial seeds by spacing the seeds rather than searching for a contiguous matching block. Special scoring matrices have been developed for specific purposes, such as PHAT and SLIM for transmembrane proteins and

```
gi|94442885|emb|CAJ28922.1|  tumor protein p53 mutant form [Homo    650    0.0
gi|38049097|gb|AAR10356.1|   tumor suppressor p53 [Homo sapiens]     648    0.0
gi|6653139|gb|AAF22640.1|    p53 tumor suppressor protein [Tupai...  640    0.0
gi|1532044|emb|CAA62216.1|   p53 protein [Oryctolagus cuniculus...   602    8e-171
gi|2440123|emb|CAA04478.1|   tumor suppressor [Marmota monax] >...   600    4e-170
gi|51480680|emb|CAH03844.1|  p53 protein [Spalax judaei]            597    3e-169
gi|18997097|gb|AAL83290.1|   P53 [Delphinapterus leucas] >gi|75...   590    3e-167
gi|75914683|gb|ABA29754.1|   p53 gamma isoform [Homo sapiens]       582    1e-164
gi|642241|emb|CAA25652.1|    p53 [Homo sapiens]                     578    2e-163
gi|94442889|emb|CAJ28923.1|  tumor protein p53 mutant form [Homo    567    4e-160
```

FIGURE 1.7: A BLAST Report

OPTIMA, which improves detection of distant homologues. The development of faster, and/or more sensitive or more flexible versions of BLAST is ongoing.

1.4 PAIRWISE LIMITATIONS

All linear pairwise alignment algorithms use some kind of scoring matrix, whether it be a simple identity scheme, with a score of one added for each match, or whether it be a well-conceived range of overall values such as the BLOSUM series of amino acid scoring matrices. These provide the highest positive scores when a residue is matched to itself, fairly high scores when a residue is matched with a very similar residue (a leucine matched to an isoleucine, for example), and negative scoring values when dissimilar residues are matched.

Gaps are allowed in most modern alignment methods, frequently with affine scoring that allows for a large scoring penalty to open a gap and a smaller one to extend it. Double-affine scoring methods allow for a third penalty when the gap gets to a certain size, allowing the alignment to cross large introns.

The problem with all pairwise alignment methods is that they treat every position in the sequence equally. This is fundamentally incorrect. Some residues are in locations where no substitution of any kind is allowable, that is, the function of the protein would be changed if a substitution were made at that location. Examples are residues in active sites, where an enzyme needs to dock, or at a crucial turn site. Other residues may be on loops that can be of various lengths. In this case, residues may be added or lost without the loss of function.

Pairwise alignment algorithms provide a generic scoring method that is useful in many cases. In some cases, however, it produces incorrect results, due to the lack of any information about the position in a particular protein or nucleotide sequence. This class of algorithm only compares one sequence to another, with no prior knowledge of other members of the protein family.

1.5 THE ADVANTAGES OF PROFILE METHODS

If you look at a multiple sequence alignment (MSA) of a family of proteins, you will notice that some locations have residues that are completely conserved across the entire alignment, and other locations that are highly variable. Some locations will have gaps across several of the family members, where others will have no gaps in any sequence. What we are seeing is the result of evolutionary progress in this family of proteins. The variable regions in the alignment are regions where mutations were allowed to accumulate without causing a loss of function in the protein. Conserved regions are sections where the protein would be 'broken' (that is, it would lose its function) if there were to be a change.

Michael Gribskov pioneered the use of the Profile method to create a scoring matrix that was position-specific. The Gribskov style profiles were trained on a set of aligned protein sequences, and would then try to capture the information found in the entire family. An implementation of the original Gribskov-style Profile search is sold commercially in the Wisconsin Package from Accelrys.

1.6   THE RISE OF PROFILE-HMMS

Profile-HMMs are commonly referred to simply as HMMs in the field of bioinformatics, and we will follow that convention here. HMMs were first proposed for use in bioinformatics by Gary Churchill of Cold Spring Harbor Laboratory. The original development of Profile Hidden Markov Models, however, took place at the University of California, Santa Cruz, in David Haussler's group. The UCSC group developed an HMM implementation called the Sequence Analysis Method (SAM). While at Santa Cruz, Kimmen Sjolander demonstrated the benefits of using dirichlet priors to improve scoring. Kevin Karplus, Christian Barrett and Richard Hughey built a program called T98 that starts with an amino acid sequence, compares it to a database, and then builds a SAM model out of the hits. This powerful program has been updated, and the latest version is referred to as T02. Meanwhile at Washington University, Sean Eddy developed a suite of Profile Hidden Markov Model tools called the HMMer package (pronounced Hammer, to indicate that it is more precise than a BLAST). This became the basis for the Protein Family database, commonly known as Pfam. Ewan Birney developed the WISE tools that extended the capabilities of HMMer models to do things like translate nucleotide sequences in all six possible reading frames. The folks at the NCBI developed a BLAST version of this idea, generating a remarkably fast and useful tool known as PSI-BLAST. We will devote more time to PSI-BLAST in a later section. Other related methods include algorithms such as Meta-MEME from the San Diego Supercomputer Center. More recent work has involved the comparison of one profile HMM to another, using algorithms such as HHsearch from Soding at the Max Planck Institute.

It has been shown by Park et al. that using HMMs instead of pairwise methods can increase the amount of distant homologues that can be detected by 300%. This fact has not been lost on the bioinformatics community, and the use of HMMs has blossomed in nearly all facets of genomic research over the past few years.

HGKKVLHA
HGKKVLHA
HGKKVAHA
HGKKVGHA
HGVTVLHA

FIGURE 1.8: A MSA

1.7 REGULAR EXPRESSIONS

There are several methods for representing protein signatures other than Hidden Markov Models. Computer scientists have long used something called 'regular expressions' to find patterns in text strings and represent ambiguous characters for a long time. While not all systems use exactly the same grammar in building regular expressions, the basic idea is the same across all platforms. The regular expression, sometimes referred to as a regexp or RE, is a structured formalism that is an efficient method for finding a set pattern in a file with a bunch of text, like a sequence database.

In a regular expression, square brackets represent choices. One of the choices will match the string. The pattern 'a[bcd]e' will therefore match 'abe,' 'ace' and 'ade' but not 'aee.'

In a regular expression a dot will match anything, so the pattern 'a.e' will match 'axe,' 'ate' or anything else with one character between the 'a' and the 'e.' If you want to use multiple repetitions of a character or pattern, a range may be entered in curly brackets, as in the pattern ab2,5c indicating that an 'a' will be followed by any number of 'b' from two to five, and then a final 'c.' This regular expression matches 'abbc' but not 'abc' because one b is not enough for the pattern. If one wants to use an unlimited number of pattern repetitions, a '*' may be used to represent zero or more of the pattern. There are many regular expression tutorials online for more extensive information– this is just a small taste of what regular expressions can do for you.

But now let us move on to a biological application of regular expressions. Let's say we have an aligned family of proteins, as in figure 1.8.

This information can be represented by a regular expression as follows:

HG[KV][KT]V[LAG]HA

The square brackets indicate that any of the characters within can be used at that position. This regular expression means an 'H' is followed by a 'G,' followed by either a 'K' or a 'V,' followed by a 'K' or a 'T,' followed by a 'V,' followed by either an 'L,' 'A' or 'G,' followed by an 'H,' then an 'A.' This regular expression may not look like much to you, but it can be used to rapidly search for matches to this pattern in a database of sequences. Prosite and PRINTS are two databases of regular expressions for bioinformatics searching.

PHI-BLAST from NCBI, eMOTIF from Doug Brutlag's group at Stanford, and FingerPrintScan from Phil Scordis are just a few of the systems for regular expression searches. In addition, virtually all computer systems now can utilize standard regular expression programs such as **grep** (which stands for General REgular exPression) for searching in text files and perl, which is an interpreted language that can produce easily ported tools for searching these types of databases. The ScanProsite tool, for example, has executables for Linux, Irix, hpux, mac, Solaris and Windows.

1.8 BUT WHAT EXACTLY *IS* AN HMM?

A group of aligned sequences from a particular family can tell us much more than a single representative sequence. For example, some of the positions in that family will be highly conserved, while others show little or no conservation whatsoever. What this shows is that those positions were important enough through evolutionary time that any change at that location caused a disruption in the function of that protein. Positions that demonstrate a high level of substitution or gapping may be considered less important functionally.

The reason behind this usually becomes clear if the proteins are modeled, but the crucial point for most purposes is not the details of why the positions are conserved, but simply to what levels, and what substitutions, if any, are commonly seen at that position. We would like to capture the information that is provided in a MSA of a protein family in a more useful way, so that we may easily see if the patterns in this family are also found in any other sequences in a database of sequences.

To see how this is done, let us first go back to Markov chains and see how they might be used for sequence analysis. A Markov chain provides a probabilistic model for a sequence family where the specific states are hidden, and each state is dependant on the one before it. A Markov chain is simply a collection of states, with a certain transition probability between each state. Any sequence can be traced through the model from one state to the next through these transitions. The model is referred to as 'hidden' because you do not know which series of transitions produced that state. An HMM that models a protein will have one state for every position in the sequence, and each of those states would have over twenty possible outcomes, one for each amino acid, and two for the insert and deletion states. A sequence may be scored by tracing it through the model by passing it from one state to the next, calculating the probabilities for each transition. Since each transition has a finite probability, the chain is more likely to emit sequences with certain outcomes at each position. HMMs may therefore be seen as position-specific profiles that represent a particular sequence family.

As a first example of an HMM, let's return to the short alignment from the last section:

HGKKVLHA

HGKKVLHA

HGKKVAHA

HGKKVGHA

HGVTVLHA

We represented these data as a regular expression of the form

HG[KV][KT]V[LAG]HA.

What is wrong with this? Take a look at the third position. We have four K's and one V, and the regular expression shows this as [KV] which means 'either K or V.' This is not strictly accurate, however. According to the training set, there is a much greater likelihood that you will see a K than a V. Our regular expression pattern does not tell us anything about probability! We can see that based on our data, we are four times more likely to have a K at position 3 than a V. We can instead represent these data as shown in figure 1.9.

This is a simple profile Hidden Markov Model—we will add more features later. This tells us not only the residues that we can expect to see at each location, but also the probability of each residue at each position. We can see that the third residue has a probability of 0.8 of being a 'K' and 0.2 probability of being a 'V.' On the other hand, at position 6 the probability of a 'G' is 0.2, the probability of an 'A' is also 0.2, and the probability of an 'L' is 0.6.

It is important to note that the probabilities are independent, and therefore any state is independent of any previous or following state. In this example, we might have a T in the 4th position. Looking at this information, we have no way of knowing what character was in the 3rd position—most likely it was a 'K,' but we can't be sure.

Higher order Markov models are possible, but are too computationally intensive to be useful for large-scale biological problems at this time. Higher

Position-	1	2	3	4	5	6	7	8
Prob(H)	1.0						1.0	
Prob(G)		1.0				0.2		
Prob(K)			0.8	0.8				
Prob(V)			0.2		1.0			
Prob(T)				0.2				
Prob(A)						0.2		1.0
Prob(L)						0.6		

FIGURE 1.9: A Sequence Profile

order Markov models take into account neighboring values into their probability scores, which might increase the overall sensitivity. This benefit has been shown to be minor, at best.

Scoring for the standard pairwise algorithms may be seen as rather generic in comparison to a Hidden Markov Model. The Pam or BLOSUM scoring matrices assign a fixed score to a particular substitution regardless of its location in the protein. That substitution, however, could have considerable effect on the function of the protein in some cases, and no effect at all in others. The commonly quoted 'percent identities' of pairwise alignments can be quite deceiving. A match with a relatively high percent identity may still be missing a crucial motif that is a characteristic signature of the protein family.

1.9   CURATED VS. NON-CURATED DATABASES

One of the troubles with using a pairwise alignment tool such as BLAST is that the large repository databases are not typically curated. This means that any mistake made upon submission tends to remain in the repository until the mistake is discovered. If, for example, someone types in a random series of amino acid characters and submits it as a 'FOOBAR' protein, it can be accepted without complaint. Any later proteins that have a significant BLAST similarity to this random sequence are likely to be submitted as 'similar to FOOBAR protein.' It is theoretically possible to produce an entire class of 'FOOBAR' protein sequences that are based on an original bogus entry.

Many of the Profile-HMM databases, such as Pfam-A and TIGRFAM, are manually curated. This means that a group of people spend a lot of time analyzing the contents of each entry, which dramatically lowers (although it does not eliminate) the possibility of spurious entries. Hidden Markov Model databases are considered secondary databases because they are derived from other databases. Primary databases, on the other hand, are those in which the data are submitted directly to the database. Primary databases include GenBank, DDBJ (DNA Database of Japan), and the EMBL Nucleic Sequence Database. Anyone can submit any sequence to a primary database and call it anything they want without any sort of quality check.

On the other hand, the community at large will help to maintain the quality of the models in the HMM databases. If, for example, the Pfam curators incorrectly include a bogus entry in the alignment for the APH_6_hur model, which is an aminoglycoside/hydroxyurea antibiotic resistance kinase, then they will very likely hear from the researchers around the world that work with this particular type of kinase. The next Pfam release will then remove the offending sequence from the alignment, which will then improve the quality of this entry.

1.10 DISADVANTAGES AND LIMITATIONS OF PROFILE-HMMS FOR BIOINFORMATICS

With the advantages that are outlined here for the use of HMMs in bioinformatic analyses, why are other methods such as BLAST still so popular? There are several limitations and disadvantages to using profile methods that keep them from becoming ubiquitous.

The most obvious issue is speed. As the HMM databases grow, the amount of time to compare each sequence to those databases grows accordingly. The increase in computer CPU clock speed is insufficient to solve this problem because the databases are growing at speeds that exceed Moore's law, so the per-sequence search time is actually growing over time.

The speed problem is actually much worse than might be predicted by the simple growth of the databases. While the 'per-sequence' search time is indeed growing somewhat, the more important factor is the way that these tools are being used. Instead of searching a few sequences at a time, it is now much more common to search thousands or tens of thousands of sequences against a database as a regular occurrence. These types of searches can typically take days of CPU time on a server, which can prove to be an extreme irritation for other users in the group. Some suggestions for ways to alleviate the time problem are given in chapter 7.

The second issue with HMM databases is coverage. There is a considerable lag between sequence identification and the time that a group of proteins is recognized as a family, or the time that a number of aligned sequence segments is recognized as a domain.

The third reason pairwise algorithms are still used is for comparative genomics. If, for example, you want to look at syntenic relationships between two related organisms, then you would want to use MegaBLAST or BLAT or some other pairwise algorithm. Not only would this be faster, but it will paint a better picture of any translocations that might be present.

1.11 SUMMARY

Pairwise alignments have progressed over the past thirty years from the global alignments and full dynamic programming methods of the Needleman-Wunsch algorithm, to the local alignment methods of the Smith-Waterman algorithm and the faster but somewhat less sensitive heuristics of FASTA and BLAST.

While all of these different types of pairwise alignment algorithms have been shown to be useful throughout the field of bioinformatics, signature methods can find more distant homologues due to the input from an entire group of sequences rather than just one.

Regular expressions have proven to be useful and powerful tools for representing motifs and using those motifs to rapidly discover families in text databases. Sequences can now be searched against collections of regular expressions in databases such as PRINTS and ProSite through the use of tools including eMOTIF, FingerPrintScan and PHI-BLAST.

Profile methods have been developed to construct a Position Specific Scoring Matrix (PSSM) based on an aligned sequence family. This PSSM is then used to score a new sequence and thus provides a comparison to the entire family rather than just to a single member of the family.

Hidden Markov models add a probabilistic component to the signature, and thus provide a better representation of the MSA data than either a profile or a regular expression. HMMs are particularly useful in discovering remote homologies in comparison to other techniques.

Some of the HMM implementations that are important in the field of bioinformatics include SAM (the Sequence Analysis Method), HMMer, the Wise Tools, and related methods such as PSI-BLAST and Meta-MEME.

1.12   QUESTIONS

1. What are some of the advantages/disadvantages of using Profile HMM type searches over pairwise alignments?

2. Print the two representations of a Protein Tyrosine Kinase in ProSite format (PS50011) and Pfam format (PF07714). Compare the two data representations and discuss the advantages and disadvantages of each format.

3. Search the following ProSite pattern: **Y - K - [DE] - [SG] - T - L - I - [IML] - Q - L - [LF] - [RHC] - D - N - [LF] - T - [LS] - W - [TANS] - [SAD]** against the Swissprot database, and report the output.

4. Obtain a Fasta protein sequence from your instructor and run it on the NCBI BLAST website. Print the output.

5. Run the sequence from problem 4 against Pfam and compare the output to what was reported by the BLAST search.

References

[1] D. Bouchaffra and J. Tan, "Protein Fold Recognition using a Structural Hidden Markov Model," Pattern Recognition, ICPR, Volume 3, 20-24 Aug. 2006, pp. 186–189.

[2] A. Krogh et al., "Predicting Transmembrane Protein Topology with a Hidden Markov Model: Application to Complete Genomes," *J. Mol. Biol.*, **305**, 567–80 (2001).

[3] E. Sonnhammer, G. Heijne, and A. Krogh, "A Hidden Markov Model for Predicting Transmembrane Helices in Protein Sequences," *Proceedings of the Sixth International Conference on Intelligent Systems for Molecular Biology*, AAAI Press, Menlo Park, CA, 1998, pp. 175–82.

[4] G. Tusnady and I. Simon, "Principles Governing Amino Acid Composition of Integral Membrane Proteins: Application to Topology Prediction," *J. Mol. Biol.*, **283**, 489–506 (1998).

[5] G. Tusnady and I. Simon, "The HMMTOP Transmembrane Topology Prediction Server," *Bioinformatics*, **17**, 849–50 (2001).

[6] K. Meln, A. Krogh, and G. von Heijne, "Reliability Measures for Membrane Protein Topology Prediction Algorithms," *J. Mol. Biol.*, **327**, 735–44 (2003).

[7] J. Zhu, J. Liu, and C. Lawrence, "Bayesian Adaptive Sequence Alignment Algorithms," *Bioinformatics* **14**, 25–39 (1998).

[8] I. Holmes and R. Durbin, "Dynamic Programming Alignment Accuracy," *J. Comput. Biol.* **5**, 493–504 (1998).

[9] P. Lio, J. L. Thorne, N. Goldman, and D. T. Jones, "Passml: Combining Evolutionary Inference and Protein Secondary Structure Prediction," *Bioinformatics* **14**, 726–733 (1999).

[10] E. Rivas and S. Eddy, "A Dynamic Programming Algorithm for RNA Structure Prediction Including Pseudoknots," *J. Mol. Biol.* **285**, 2053–2068 (1999).

[11] C. Burge and S. Karlin, "Prediction of Complete Gene Structures in Human Genomic DNA," *J. Mol. Biol.* **268**, 78–94 (1997).

[12] D. Kulp, D. Haussler, M. G. Reese, and F. H. Eeckman, "A Generalized Hidden Markov Model for the Recognition of Human Genes in DNA," *Proceedings of the Fourth International Conference on Intelligent Systems for Molecular Biology*, AAAI Press, Menlo Park, CA, 1996, pp. 134–142.

Chapter 2

Profile HMM Models

2.1 INTRODUCTION

Chapter 2 continues and expands the instruction on HMMs and the various states that may be found in an HMM. Chapter 2 introduces the various models, such as the 'Plan 7' model, local versus global scoring, and the maximum entropy model. While this book will primarily discuss the use of HMMs for the discovery of protein family homology, we will cover what other types of applications HMMs have in the field of biology in chapter 8.

2.2 A GENERAL MODEL FOR HMMS

In chapter 1, we looked at an alignment of five sequences that were eight characters long:

HGKKVLHA

HGKKVLHA

HGKKVAHA

HGKKVGHA

HGVTVLHA

We then looked at a profile of these sequences to calculate the probabilities at each location in the alignment. We may display this graphically as in figure 2.1.

In this figure, the M stands for the match probability at positions one through eight. For example, at M6 there will be a small probability of finding an 'A', and a larger probability of finding an 'L'.

Let us compare another sequence to this alignment:

HGVTGTGVLHA

Now we have some additional amino acids to contend with. The beginning and end of this sequence match our model, but there are three characters in

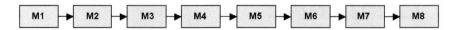

FIGURE 2.1:   Match States

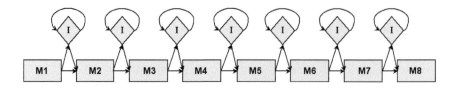

FIGURE 2.2:   Sequences with Added Data

the middle (GTG) that don't seem to fit. Our model can treat these with the same probabilistic method that it treats any other character, if we add some insertion states to extend it, as in figure 2.2.

Similarly, if we look at a sequence such as

HGVLHA

we see that this is again a good fit to the model, but with a couple of characters missing or 'deleted' from the alignment. We would like to be able to categorize the probability of an insertion or deletion at any location in the model. Just as with amino acid probabilities, there are some locations in a protein that are very resistant to insertions or deletions, and other locations that are very amenable. For example, an active site of an enzyme would be very resistant to insertions or deletions, because this is the region that binds the substrate. Any insertion or deletion at this point will likely reduce the binding energy so much that the enzyme will no longer function properly.

On the other hand, insertions or deletions in some other regions, such as some loop structures, may be well tolerated with few or no alterations to the functionality of the resulting protein.

Adding insertion and deletion states to our model yields a more extensive system that looks like figure 2.3.

So we can see that the sequence HGVLHA would fit this model as seen in figure 2.4.

The sequence matches the model with character deletions at the third and fourth positions. We also may want to consider a sequence such as

HGKKVLHAAGAGTGHGVTVLHA

This sequence has the above pattern represented twice with some junk characters in the middle. To find both of these matches it would be beneficial

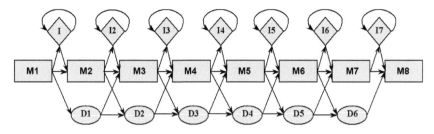

FIGURE 2.3: Match States with Insertions and Deletions

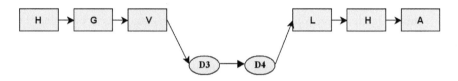

FIGURE 2.4: One Possible Path Through the Model

to provide a means to 'reload' our model to be able to detect all the instances of the pattern in the sequence, not just the first. We will also add an additional state to account for any additional characters between one instance of the domain and another.

In figure 2.5, we have an example of a shorter four position HMM so that we may be better able to fit it on the page. The B and E states represent the beginning and end, respectively. Notice that the model may be reloaded by the arrow going from the end to the beginning. In a chromosome, for example, a single domain may be found dozens of times with any amount of data in between them. The J state has a loop indicating that it may be found numerous times. These junk data, here defined as data that do not

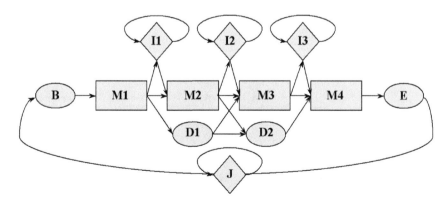

FIGURE 2.5: Adding the J State to the Model

match our model, may have any length, so we allow the J state to repeat indefinitely. The insert states also have loops indicating that there may be multiple character insertions at any location.

2.3   PLAN 7 FROM JANELIA FARMS

The plan 7 model is a representation of the system used by the popular HMMer (pronounced hammer, as in 'more precise than a BLAST') program suite. Originally, this model had nine transitions, and was to be called the plan 9 model, as in the movie, *Plan 9 from Outer Space*. However, two transitions were removed, due to the fact that one should never really see transitions from an insert state to a delete state, or vice versa.

Imagine for a moment that we are searching a long protein with the short pattern in figure 2.6. The S and T states are the start and terminate positions, respectively. The N and C states allow for sequence data before and after the match, respectively. The B and E states stand for the beginning and the end of the alignment. As with our previous examples, the M or match state indicates the probability that a particular residue will match at that location. The D and I states represent the deletion and insertion probabilities, respectively. The J state shown at the bottom allows the model to match the sequence multiple times. This state allows any amount of sequence data of any kind ('junk' data, hence the 'J' state). This is useful when the model represents a protein domain and that domain is found several times in the protein, or when the query is an entire chromosome and the protein family is found several times in that chromosome.

Notice the dashed lines—these indicate that the model can go straight from the beginning to any point in the model. Similarly, each match state in the model may go directly to the exit state. These dashed lines allow for the model to be local in nature. If the probabilities for these dashed-line relationships

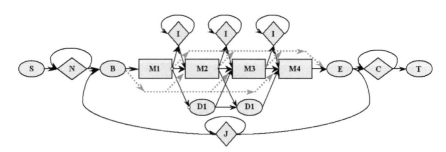

FIGURE 2.6:   The Plan 7 Representation of a Short, 4 Position HMM

were to be set at zero, then any hit to the model would have to be fit to the entire model, from beginning to end, forcing a global alignment.

As an example, let us say that we have a model with nine match states and a consensus sequence as follows:

AGRANDMAN

and we align it with the sequence

GRANDMA

We can see that this will jump directly from the beginning B state to the second match state, or M2. Then again at the end, we jump from M8 to the exit state, skipping the last match state.

2.4   LOCAL SCORING

Scoring the comparisons of these models may be done at either the global or the local level. Local scoring allows the alignment to start anywhere in the model, and end anywhere in the model. This frequently makes sense, in that the entire model does not always need to be found in the sequence to indicate the presence of the protein family. Fragments of domains may be found in EST collections and truncated sequences, or they may occur when a domain is inserted into an existing domain.

On the other hand, one must be wary of claiming that a short hit represents a domain or family—as an example, I once saw someone report a 'match' that was only one character long! Used wisely, however, local alignments can provide useful results. The Pfam-fs database is set up with local scoring models, as is TIGRfam-frag.

2.5   GLOBAL ALIGNMENTS

It is important to note here that by 'global alignments' I mean global with respect to the model, not the sequence. This is what is usually referred to as 'glocal' in the Pfam database. The model may be found several times in the sequence, but only if the entire model is found each time.

Pfam-ls contains models that are built with a glocal parameter (hmmbuild –F hmm_ls). This sets the beginning-to-match transitions and the match-to-end transitions to zero probability for all but the first and last match states,

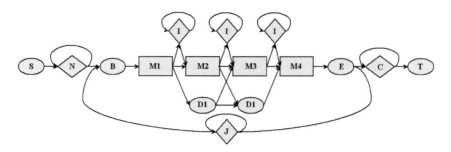

FIGURE 2.7:   The Dashed Lines Are Effectively Erased in the Glocal Mode

respectively. Another way to look at it is that you are getting rid of the 'dotted-line' transition probabilities as shown in figure 2.7.

Glocal mode assumes that the start and end points of the model are absolutely correct, and in some cases this can lead to incorrect results on the edges of the domain. Genewise has a really nice solution to the global vs. local problem that is called –wing. The Wing option provides a local mode for the first fifteen model positions and the last fifteen model positions, but forces a global alignment in the central part of the model.

Pfam searches are usually done in a merged method, searching both Pfam-ls and Pfam-fs and then reporting the best hit.

True global alignments would be used if the model and the sequence are of similar length. Global scoring forces the beginning of the model to match the beginning of the sequence, and the transitions along the path generate the sequence until the end of the model reaches the end of the sequence. This method is of limited use.

2.5.1   Calculating Transition Probabilities

When a protein family is aligned, there are regions that are not possible to align well, and many insertions will be seen in the MSA. There are many reasons why this may occur, for example, the proteins may not have the identical structure in the loops, or the lengths of the loops may vary considerably between one protein and the next. In this situation the areas of conserved secondary structure are aligned well, but the loop regions are not.

The scoring of the well-aligned areas will reflect relatively high scores for the match states, but very poor scores for the insert or delete states. On the other hand, the regions of poor alignment will show higher scores for the insert states, and that probability will increase with the number of sequences with inserts at that position. The longer the poorly-aligned region is, the higher the transition probability from an insert state to itself. Similarly, the transition to the delete state will have a higher probability as the number of sequences with no residues in that position increases.

When assigning the transition probabilities, it is important to account for background rates—that is, the possibility that there may be transitions that are not observed in the training alignment. This possibility leads us to the use of prior information. Prior information about protein sequences in general can help us to construct a model that can more accurately recognize remote homologs than a model that only has data from the training set. No transitions should be set with a zero probability. For example, insertions and deletions are possible anywhere in a sequence, even if we have not previously observed one at that location. Using prior knowledge of insertion and deletion rates from a large number of sequences can help us to set the background rates. These rates are then combined with information from the alignment. The probabilities are set by the ratio of the background information to the alignment information. This ratio is dependent upon the significance of the observation, which is based on the number of sequences in the alignment. Ultimately, no transition probability should be set to zero.

2.5.2 State Probabilities

There is some probability, however small, that any amino acid may be found at any location in a protein. If a model is to be capable of matching new and distant members of the family, it will have to take this into account. Therefore, the model will have to allow for the possibility of amino acids being found in positions where none are found in the training alignment. These probabilities may be derived from standard matrices such as the BLOSUM series, which gives the likelihood of one amino acid being substituted for another. In general, these probabilities are dependent upon properties such as relative size, hydrophobicity and charge. The substitution matrix is typically used to generate pseudo-counts based on the amino acids observed at each location in the alignment.

The BLOSUM62 scoring matrix shown in figure 2.8 depicts the observed substitutions found in a wide sampling from the aligned segments of

	C	S	T	P	A	G	N	D	E	...
C	9	−1	−1	−3	0	−3	−3	−3	−4	
S	−1	4	1	−1	1	0	1	0	0	
T	−1	1	4	1	−1	1	0	1	0	
P	−3	−1	1	7	−1	−2	−1	−1	−1	
A	0	1	−1	−1	4	0	−1	−2	−1	
G	−3	0	1	−2	0	6	−2	−1	−2	
N	−3	1	0	−2	−2	0	6	1	0	
D	−3	0	1	−1	−2	−1	1	6	2	
E	−4	0	0	−1	−1	−2	0	2	5	
...										

FIGURE 2.8: A Portion of the BLOSUM62 Matrix

polypeptides. This gives us a method of estimating probabilities for residues that have not appeared in the initial training alignment.

Let us examine what this means for our example alignment:

HGKKVLHA

HGKKVLHA

HGKKVAHA

HGKKVGHA

HGVTVLHA

In the first position, H is 100% conserved. Looking at the BLOSUM62 matrix, we can see that H gets a positive score of 8 when matched to itself, but most everything else is negative or zero. The exception is Y. We have not seen any substitutions in the first position of this alignment, but if there were it would most likely be a Y. Less likely would be E, Q and R, and least likely of all would be C, I, L and V, based on the scores in the BLOSUM62 matrix.

2.5.3   Building a Better Model

When building a model, it is best to start with a large group of related sequences that are as diverse as possible. When these sequences are aligned, typically there will be some groups of closely related sequences that are more distantly related to other groups of genes. If we simply give equal weight to each sequence when assigning probabilities to the model, then the largest subgroup will have the largest representation. The resulting model will not properly represent the true diversity of the family, and will not be very useful at finding distantly related members. Weighting the sequences will capture the information in the subgroups while retaining the valuable information found in the differences between closely related members of the subgroups.

The scheme that is used by most systems is the maximum entropy model. This involves making the statistical spread of the model as broad as possible. This creates a more general model, and the model will then be best able to find remote members of the family.

2.6   THE MAXIMUM ENTROPY MODEL

As with the concept of Hidden Markov Models themselves, the concept of maximum entropy is borrowed from other branches of information theory. Maximum entropy modeling is currently used in fields such as computer vision, natural language processing, spatial physics and several other areas. Natural language processing, in particular, has a good deal of interest to the bioinformaticist, as it provides a means for data mining scientific papers for information.

Back in 1957, Edwin Jaynes wrote about establishing probability distributions using something called the maximum entropy estimate, which he described as a type of statistical inference. Jaynes stated: "It is the least biased estimate possible on the given information; i.e., it is maximally noncommittal with regard to missing information."

What this means is that if we want to characterize some unknown events (in our case a family of biological sequences) with a statistical model, we should always choose the model that has the maximum entropy, if we wish to avoid a bias about missing data.

What does this imply for those of us who want to model protein family data? The term 'maximally noncommittal with regard to missing information' simply means that we don't want to assume that we will never see a particular residue at a particular position, just because we have not seen it in the training set. Being 'maximally noncommittal' means that there is always some possibility that any residue may be found at any location. In practice, that probability may be very low, but we cannot rule it out completely.

One practice that has become more prevalent in recent years is that of making several more specific models, instead of one more general model to represent a protein family. The TLFAM database is an example of this trend–models in this database (actually a set of databases) are trained on a specific class of organisms. Therefore, TLFAM-pro is designed for use in prokaryotes; TLFAM-arc is designed for the study of archaeons, and so forth. Other database systems, such as PANTHER(figure 2.9), break each protein family into a set of derivative subfamilies by functional similarity.

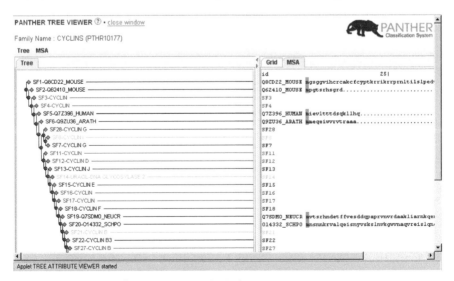

FIGURE 2.9: PANTHER Family Information

Family:	CYCLINS (PTHR10177)
Subfamilies:	24
PANTHER Links:	
PANTHER Molecular Function:	Molecular function unclassified
PANTHER Biological Process:	Cell cycle
Cell cycle control	
Pathway Categories:	No pathway information available
Training Sequences:	295
HMM Length	360
Downloads:	HMM (HMMER format)

Genes Assigned to this Family

	Total	Celera	FlyBase	NCBI
H. sapiens	41	22	0	19
M. musculus	42	24	0	18
R. norvegicus	34	17	0	17
D. melanogaster	7	0	7	0

FIGURE 2.10:   PANTHER Information on the Cyclins Family

It might seem at first that these methods violate the rule of maximum entropy, but this is not the case. Breaking the protein families down by either one of these methods ultimately captures more of the information than combining them all into a single model. The Cyclin family, for example, has 24 subfamilies in the PANTHER database (figure 2.10). As a single model, much information would be combined, with an inevitable loss of detail, as this family is involved in a wealth of functions. Searching at the subfamily level provides the overall annotation that is needed to predict the specific function of the sequence of interest.

2.7   STATISTICS

One cannot typically determine the quality or usefulness of an alignment simply by the score, the number of matches, or the number or percentage of similarities in the alignment. What is more useful is a way to rate the significance of that alignment, in a comparison with other alignments. Karlin and Altschul proposed such a method in 1990, and this has become the foundation of many software implementations. HMMer, SAM, BLAST and many other bioinformatics software packages use E-value, or expectation value. What we would like to know is this: what is the likelihood that a partic-

```
Pkinase_Tyr: domain 1 of 1, from 481 to 560: score 111.7, E = 3.1e-32
                *->lklgkkLGeGaFGeVykGtlkg...sgegtkikVAVKtLkeigasse
                   l+lgk LGeG+FG V+++++ g ++++ ++ ++VAVK+Lk+ +a+++
      pFGFR2   481    LTLGKPLGEGCFGQVVMAEAVGidkDKPKEAVTVAVKMLKD-DATEK 526

                   eieredFlrEAsiMkklGdHpNiVrLlGvctkegePggpgl<-*
                   +  + d+ +E+++Mk +G+H+Ni++LlG+ct+ ++P    l
      pFGFR2   527 D--LSDLVSEMEMMKMIGKHKNIINLLGACTQ-DGP----L    560
```

FIGURE 2.11: An Example Alignment

ular alignment was found by random chance? After all, with the rather large databases in use today, it is fairly likely that a match of some quality would be found with virtually any sequence that could be typed in by a cat walking across the keyboard (assuming that the non-amino characters were filtered out)!

We should also note that the likelihood of finding a hit by random chance is going to increase linearly with the size of the database. So, what the expectation value provides is an estimate of the number of alignments of a particular score that would be likely to be found by random chance.

In figure 2.11 we have an alignment of a sequence called pFGFR2 to the tyrosine kinase domain from the Pfam database. The bottom line represents the sequence, the top line indicates the emission states from the model.

2.8 OTHER USES FOR HMMS IN BIOLOGY

The primary purpose of this book is to discuss the application of Profile-HMMs to the problem of sequence analysis. It is important to note, however, that the biological applications go far beyond these applications. They can also be used for

- detection of CpG islands

- signaling region identification

- prediction of trans-membrane helices

- gene prediction

- protein/protein interactions

- protein/ligand binding site predictions

- post-translational modification predictions

Some of these uses discussed in more detail in chapter 8.

2.9 SUMMARY

We have looked at the general model of HMMs to gain a greater under-standing of the way these tools can be used to represent a set of training data. We have seen how these models represent amino acid frequencies and how the models can handle insertions and deletions. We also covered the importance of the J state for multidomain protein sequences.

The 'Plan 7' model from the HMMer package by Sean Eddy is discussed here because of the predominance of this package in the field. HMMer is provided as an open-source system, and has therefore been ported to a large number of applications, both commercial and academic. Several parallel versions of HMMer exist, in both PVM and MPI versions. HMMer has also been accelerated by several companies, both in software and hardware.

Local scoring is a term that indicates that any length of the query can match any length or fraction of the target. Separate versions of many HMM databases are available to support this method of comparison. This mode should be used wisely, as a match to a small part of a protein domain will not necessarily mean that the protein domain is actually present.

The other commonly used search format is the so-called 'glocal' mode. This ensures that the entire profile HMM is found within the hit, but allows the hit to be local with respect to the sequence. The glocal mode is enabled by setting the transition probabilities from any position other than the final one to the exit state to be equal to zero.

True global alignment mode in an HMM search would be of limited use. The sequences would have to be known to be complete proteins or complete domains, and then only those models representing that type would find a reasonable match. That is, the domains would not be found in a multidomain protein because the alignments for each domain would be forced to match the entire length of the sequence.

The maximum entropy model states that the best model to represent a protein family is one that maximizes the entropy or diversity of the training set. If this model is followed, the resulting models will have the best sensitivity towards distantly related sequences. A common problem amongst those beginning to model proteins in this way is to gather sequences that are too similar to one another, providing an over-representation of the initial seed sequence.

Statistical results are typically used as a cutoff threshold because they provide an understanding of how 'real' the match is. The expectation value is an estimation of the chance that a hit of this quality would be found by a random sequence. Therefore, an alignment with an E-value of 10^{-3} would have one chance in a thousand of being found by a random sequence.

2.10 QUESTIONS

1. What might be the consequences of searching a model against a translated chromosome with the J state disabled?

2. What would be the result of a global alignment between an EST sequence and a chromosome?

3. The website for the PANTHER database is www.pantherdb.org. What are the molecular functions, biological process and pathway of the PTHR10026 cyclin family?

4. The Plan 7 model eliminates insert-delete transitions as well as delete-insert transitions. Why are these not necessary?

5. Run a set of sequences provided by your instructor against the uniref90 database at www.uniprot.org. Run these same sequences against the Pfam database at any Pfam search site. Compare and contrast the progression of the E-values as you progress from the best hit to the worst.

References

[1] A. Krogh, "Two Methods for Improving Performance of a HMM and Their Application for Gene Finding," *Proceedings of the Fifth International Conference on Intelligent Systems for Molecular Biology*, T. Gaasterland, P. Karp, K. Karplus, C. Ouzounis, C. Sander, and A. Valencia, Eds., AAAI Press, Menlo Park, CA, 1997, pp. 179–186.

[2] E. Birney and R. Durbin, "Dynamite: A Flexible Code Generating Language for Dynamic Programming Methods Used in Sequence Comparison," *Proceedings of the Fifth International Conference on Intelligent Systems for Molecular Biology*, T. Gaasterland, P. Karp, K. Karplus, C. Ouzounis, C. Sander, and A. Valencia, Eds., AAAI Press, Menlo Park, CA, 1997, pp. 56–64.

[3] E. Birney, "Sequence Alignment in bioinformatics," Ph.D. thesis, The Sanger Centre, Cambridge, U.K., 2000; available from ftp://ftp.sanger.ac.uk/pub/birney/thesis/.

[4] A. Krogh, M. Brown, I. S. Mian, K. Sjlander, and D. Haussler, "Hidden Markov Models in Computational Biology: Applications to Protein Modeling," *J. Mol. Biol.* **235**, 1501–1531 (1994).

[5] S. R. Eddy, "HMMER: A Profile Hidden Markov Modelling Package," available from http://hmmer.wustl.edu/.

[6] A. Bateman, E. Birney, R. Durbin, S. R. Eddy, K. L. Howe, and E. L. L. Sonnhammer, "The Pfam Protein Families Database," *Nucleic Acids Res.* **28**, 263–266 (2000).

[7] J. Schultz, R. R. Copley, T. Doerks, C. P. Ponting, and P. Bork, "SMART: A Web-Based Tool for the Study of Genetically Mobile Domains," *Nucleic Acids Res.* **28**, 231–234 (2000).

[8] K. Sjlander, K. Karplus, M. Brown, R. Hughey, A. Krogh, I. S. Mian, and D. Haussler, "Dirichlet Mixtures: A Method for Improved Detection of Weak but Significant Protein Sequence Homology," *Comput. Appl. Biosci.* **12**, 327–345 (1996).

[9] S. R. Eddy, G. J. Mitchison, and R. Durbin, "Maximum Discrimination Hidden Markov Models of Sequence Consensus," *J. Comput. Biol.* **2**, 9–23 (1995).

[10] T. Jaakkola, M. Diekhans, and D. Haussler, "A Discriminative Framework for Detecting Remote Protein Homologies," *Comput. Biol.* **7**, 95–114 (2000).

Chapter 3

HMM Methods

3.1 INTRODUCTION

Chapter 3 presents some of the different types of HMM methods that are currently available, with the advantages and disadvantages of each one. Commercial methods, typically optimized or accelerated versions of the open-source code, are covered in more detail in chapter 7.

3.1.1 What Exactly Do You Want to Do?

In this book we will be focusing primarily on the HMMer system from Sean Eddy at Washington University. This is simply because it is an open-source system, which has led to many customizations and optimizations. HMMer is now the most widely used system for Profile-HMM analysis in bioinformatics and most databases support this format.

Within the HMMer package, there are several programs for your specific purpose. If you have a single model and wish to find that family or domain in a large flatfile database of sequences, then use hmmsearch. On the other hand, if you have a group of sequences that you would like to run against a database of HMMs, then you will use hmmPfam. This is perhaps the most commonly used program in the HMMer package, and the one that takes the most time. If you frequently have a need to analyze thousands of sequences with thousands of models, then you need to consider spreading the problem over a cluster, using optimized software such as SledgeHMMer, ClawHMMer, LDhmmer, or SPSPfam. If this is still insufficient, or if your cluster is already oversubscribed, then an accelerator is called for.

3.2 THE HMMER SUITE OF PROGRAMS

There are several programs available that use Hidden Markov Models or related profile methods. The HMMer package (which may be downloaded from hmmer.janelia.org) is the one that is most commonly used. The source

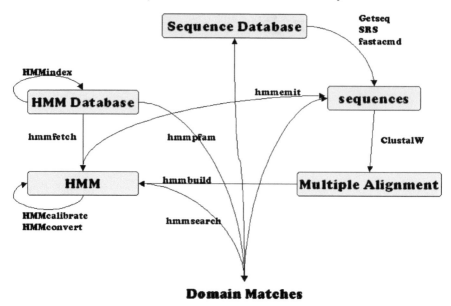

FIGURE 3.1:   HMMer Programs

code and documentation are available along with executables for a wide range of UNIX-like operating systems. For Windows ports of the HMMer package, or other variations such as parallelized versions, a bit of Internet searching may be required.

HMMer is included as a part of several commercially available software packages, such as the Wisconsin Package (commonly known as GCG), iNquiry and many others. Free software packages such as EMBOSS also include the HMMer package.

Sean Eddy, from the Howard Hughes Medical Institute Janelia Farm campus, created and maintains the HMMer package, with ports to various systems being contributed by various others. It is released under the terms of the GPL (Gnu Public License).

Figure 3.1 shows a graphical view of the various components and programs that may be used with the HMMer package. Not everyone will use all of these tools all the time, but we will cover all of them in order to give a complete picture of how they are all integrated together, and what they all do. While other packages, such as SAM, have different programs to do most of these tasks, the concepts are the same.

If we wish to build Profile-HMMs of our own, we will likely start with a sequence database. If the database is online, then we will use the provided tools for that particular server. For example, if we are pulling the data out of NCBI, then we will use Entrez. If, on the other hand, we are using EMBL or DDBJ data, then we will use SRS (Sequence Retrieval System). While

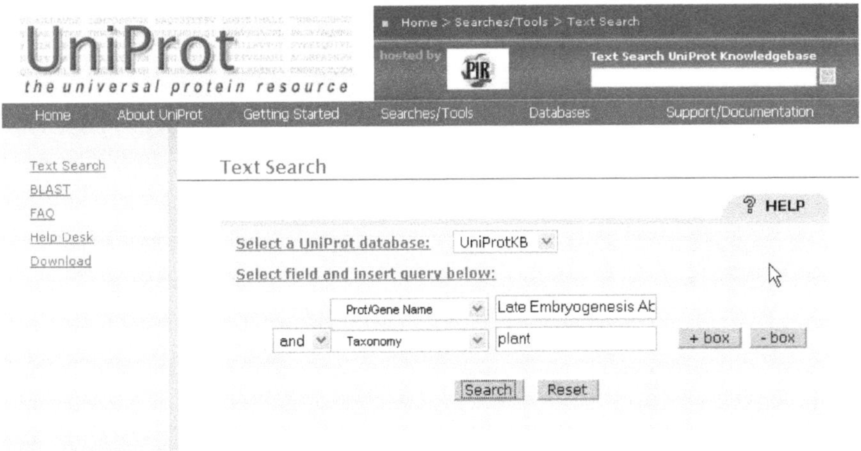

FIGURE 3.2:   UniProt Text Search

links to Uniprot data can be extracted from any of these sources, the Uniprot server has its own retrieval system, which is very powerful and flexible.

In figure 3.2, we see an example of a text search of the UniProt knowledge base. This database is derived from Swiss-Prot, TrEMBL and PIR-PSD protein sequences with added annotations of sequence and functional information. UniProtKB/SP is smaller, yet better annotated and manually curated, while UniProtKB/TrEMBL is larger and automatically machine annotated.

The text search shown in figure 3.2 demonstrates some of the power and flexibility of the interface. Over thirty fields may be searched for sequence identifiers, function, classification, properties, keywords, comments, organisms or literature publications. In this example, I searched for sequences with a protein name of 'Late Embryogenesis Abundant,' and then further restricted the search to plants. Selecting the '+ box' would add another field that would make the search even more selective.

Figure 3.3 shows the top results of the search. We have found 121 entries that meet our search criteria, with only the top two shown. If we would like to change which fields are shown in this view, we may alter them by clicking the 'Display Options' button and adding or subtracting any fields that we choose. We can BLAST individual sequences against any of the UniProt databases, which may be very useful in finding sequences that are related to our interests but are not yet annotated in a way that would bring them up in the initial search. In this example, this would involve sequences that may be Late Embryogenesis Abundant proteins that are not annotated as such. Having the BLAST capability is very handy, but many times it is too limited. BLASTing these sequences one at a time gets very tedious when you have many to do—121 in this example.

UniProt provides a ClustalW service with an interactive tree viewer that can be very useful in selecting and visualizing data, as shown in figure 3.4.

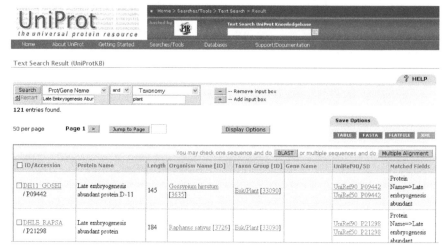

FIGURE 3.3:   UniProt Text Search Results

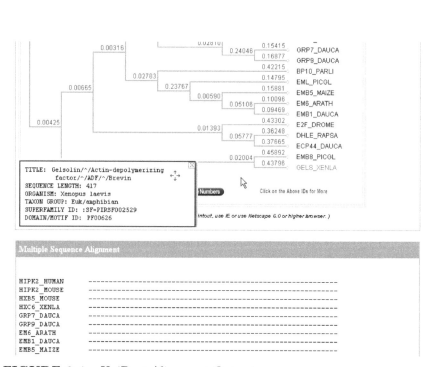

FIGURE 3.4:   UniProt Alignment Output

>DRPD_CRAPL

MAQLMNKAKNFVAEKVANVEKPKASVEDVDLK...

>LEA14_ARATH

MASLLDKAKDFVADKLTAIPKPEGSVTDVDLK...

>LEA14_GOSHI

MSQLLEKAKDFVVDKVANIKKPEASVSDVDLK...

>LEA14_SOYBN

MSQLLDKAKNYVAEKVTNMPKPEASVTDVDFK...

>LEA2R_ARATH

MASADEKVVEEKASVISSLLDKAKGFFAEKLA...

>O80576_ARATH

MSTSEDKPEIISRVVHQEGDVEIVDRSQKDKD...

>Q9SPJ4_CAPAN

MADLMEKAKNYVVEKVGNMEKPEAEITDVDMK...

>Q9ZQW6_PSEMZ

MASLMDKAKQFVVDKIAHIEKPSADVTDIDMK...

FIGURE 3.5: Downloaded Sequences (LEA-2 Proteins in This Example)

Additionally we may select several (or all) of these sequences and align them with ClustalW. Along with the alignment, a guide tree is displayed, with or without the distance values. Selecting any of the gene identifiers will bring up an information box with more data about that gene (figure 3.4) and the sequences themselves may be downloaded in FASTA format, as in figure 3.5.

Occasionally we may start with a database that resides on a local server. To extract the desired sequences from this file, we may use any number of programs, such as entret or seqret from the EMBOSS package (emboss.sourceforge.net), the getseq Perl module from BioPerl (bioPerl.org), or any of several other programs. If this function is performed often enough, it may become worthwhile to install and run SRS locally (http://www. biowisdom.com/solutions/srs/).

Once the related sequences are extracted, as seen in figure 3.5, they may be aligned with a MSA program such as ClustalW (http://www.ebi.ac.uk/clustalw/). ClustalW may be run locally through a command line interface, or through any of a number of websites. Let us assume that we have extracted a number of sequences into a single file that we shall name seqname.fa. The commands for running ClustalW locally are

ClustalW seqname.faa

This will align the sequences in the seqname.faa file into a MSA, as shown in figure 3.6. The output files will be titled seqname.aln and seqname.dnd. Information about the alignment may be obtained from the alistat program.

```
DRPD_CRAPL      ───────────────────MAQLMNKAKNFVAEKVANVE–
Q9SPJ4_CAPAN    ───────────────────MADLMEKAKNYVVEKVGNME–
LEA14_SOYBN     ───────────────────MSQLLDKAKNYVAEKVTNMP–
LEA14_GOSHI     ───────────────────MSQLLEKAKDFVVDKVANIK–
LEA14_ARATH     ───────────────────MASLLDKAKDFVADKLTAIP–
LEA2R_ARATH     ───────────────MASADEKVVEEKASVISSLLDKAKGFFAEKLANIP–
Q9ZQW6_PSEMZ    ───────────────────MASLMDKAKQFVVDKIAHIE–
O80576_ARATH MSTSEDKPEIISRVVHQEGDVEIVDRSQKDKDEEKEEGKGGFLDKVKDFI-
HDIGEKLEGT
. :::*.* :. : :
DRPD_CRAPL           ──-KPKASVEDVDLKDVGRHGITYLTRICVENPYSASIPVGE-
IKYTLKSAGRVIVSGNI
Q9SPJ4_CAPAN         ──-KPEAEITDVDMKKVSMDSISYHANVAVKNPYSVPVPIMQISYAL-
KCSGRIIVSGTI
LEA14_SOYBN          ──-KPEASVTDVDFKRVSRDSVEYLAKVSVSNPYSTPIPICEIKYSLKS-
AGKEIASGTI
LEA14_GOSHI          ──-KPEASVSDVDLKHVSRECVEYGAKVSVSNPYSHSIPICEISYNFRS-
AGRGIASGTI
LEA14_ARATH          ──-KPEGSVTDVDLKDVNRDSVEYLAKVSVTNPYSHSIPICEISFTFH-
SAGREIGKGKI
LEA2R_ARATH          ──-TPEATVDDVDFKGVTRDGVDYHAKVSVKNPYSQSIPICQISYIL-
KSATRTIASGTI
Q9ZQW6_PSEMZ         ──-KPSADVTDIDMKNLTTDSVTLESAIDITNPYDHDLPIWEISFR-
LRSADKLIASGTI
O80576_ARATH    IGFGKPTADVSAIHIPKINLERADIVVDVLVKNPNPVPIPLIDVNYLVESD-
GRKLVSGLI
.* . : .::: . : : ** :*: ::.: ... : : .* *
```

FIGURE 3.6:   Aligned Group 2 LEA Proteins

Alistat will report the alignment length, the range of sequence lengths and other useful information:

Alistat seqname.dnd

Many schemes exist to highlight the related characters in a MSA. Most of these are related to the physical and chemical properties of the amino acids in the column. Residues with similar properties tend to be substituted most easily through evolutionary processes. For example, leucine and isoleucine are both non-polar and aliphatic as shown in figure 3.7, and so may be substituted rather easily. Most coloring schemes will therefore group these two together.

The original and simplest scheme is to place characters under each column as shown in figure 3.5. Showing a "*" character under an aligned column indicates that the position is entirely conserved—that is, the residue is identical for each sequence at that position. A position with ":" under the column indicates that only conserved substitutions have been observed in that column.

Conserved residues are

AVFPMILWY (Small and hydrophobic)

DE (Acidic)

RHK (Basic)

STYHCNGQ (hydroxylic, amine and basic)

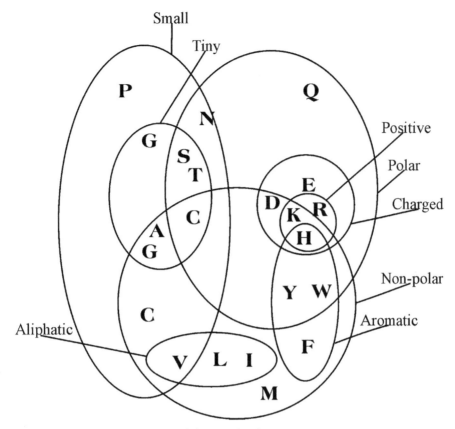

FIGURE 3.7: Properties of Amino Acids

If you wish to view the trees generated by these data, there are several useful programs such as treeview (http://taxonomy.zoology.gla.ac.uk/rod/treeview. html). Treeview is a graphical program and does not have a command line interface. Many other tree viewers are available, such as ATV (A Tree Viewer, available at http://www.phylogenomics.us/atv/), which comes as part of the Forester phylogenomics package (http://sourceforge.net/projects/ forester-atv/), and an example of ATV output may be seen in figure 3.8.

Once the alignments are created, they may be built into HMMs through the hmmbuild program using the following command:

Hmmbuild seqname.hmm seqname.aln

The statistical scores in these freshly created HMMs are not as accurate as they could be, because they have only 'learned' from a few sequences. To be able to adequately discriminate between a real signal and random data, the scores must be adjusted to obtain proper cutoff values. In order to improve

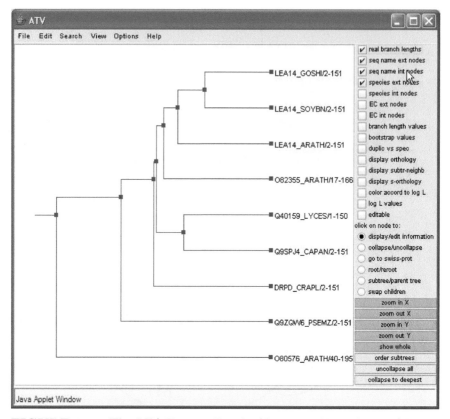

FIGURE 3.8:   The LEA Protein Family Alignment Viewed with ATV

these values we must run hmmcalibrate to run enough random data through the model to set the statistics proPerly. Hmmcalibrate takes the scores from the comparisons of the random data and the model and fits an extreme value distribution (EVD) to a histogram of those scores. The HMM is then resaved with those EVD scores incorporated in the model.

Please be aware that running hmmcalibrate takes a few minutes for each sequence (depending on the number and type of the CPUs that you are running). If you need to calibrate a large number of models, it is probably best to create a concatenated database of models and then let it run overnight, or over a weekend. The actual run time will depend on the average length of the models, the total number of models, and the speed of your computer.

The command for hmmcalibrate is simply

Hmmcalibrate seqname.hmm

The calibrated model is saved with the same name as the source file.

The HMM that you have just aligned, built and calibrated can now be used as is, or converted into one of a few different formats using hmmconvert. This program is typically used to convert models into GCG profile format, or into a binary file. An HMM in binary format is not human readable, but is considerably smaller than the default ASCII format. While this conversion is not necessary, it makes it much faster to run the search, because the time to load the database from the hard disk is greatly reduced. The command for hmmconvert is

Hmmconvert –b seqname.hmm seqname.hmm

The model may now be run against a fasta format sequence database using the hmmsearch program. I recommend that the first search that you do is to run against the database that you started with. This way you can collect any members of the family that were unannotated in your first extraction process. As an example, let us say that we are building a model of Group1 LEA proteins. We start by extracting all the proteins that are annotated as group 1 LEAs, aligning them with ClustalW and building them into Profile-HMMs with hmmbuild and calibrating with hmmcalibrate. Running this model against the initial database is likely to find a number of proteins that are annotated as 'hypothetical proteins' or some other such designation. These are possible Group 1 LEAs that have simply not yet been annotated as such. If you feel confident that these new sequences really do represent the proteins of interest, then you may add them to the existing sequences, align them all with ClustalW and rebuild and recalibrate the models.

Let's say that we extracted the original seqname.fa sequences from the uniprot database. We can then run our resulting model against our downloaded uniprot database with the hmmsearch command:

Hmmsearch seqname.hmm uniprot > seqname.out

Depending on the size of the model, the size of the database, and the type of computer that you are using, this search could take several minutes. The last part of the above command redirects the output into a file that is named seqname.out. This enables us to save the output, rather than simply watch it run past on the screen.

The Pfam database is an expertly curated collection of Hidden Markov Models that can provide an excellent method for the analysis of newly sequenced data. Given that we have a collection of unannotated protein sequences titled protseqs.fa, we may use hmmPfam to see what similarities there may be:

HmmPfam Pfam_fs protseqs.fa > Prot_vs_Pfam.out

Depending on the size and number of the sequences, the size of the database, and the type of computer that you are using, this search could take several minutes to several hours, and produce a large amount of output.

Occasionally, applications arise where one might like to simulate sequences that are consistent with the consensus of a sequence family. Hmmemit will generate these sequences in fasta format, ten at a time by default:

Hmmemit –o seqnameemit.fa seqname.hmm

Once you have gone through the alignment, building and calibration process with a number of models, they may be concatenated into a single file to make the database. This may be done using the cat command in Linux, or the copy command in Windows. An index of the database is created with the hmmindex command, which then allows you to search and extract individual models as needed with the hmmfetch program.

In Windows:

Copy seqname1.hmm+seqname2.hmm+seqname3.hmm
+seqname4.hmm allseqs.hmm

In Linux:

Cat seqname1.hmm seqname2.hmm seqname3.hmm seqname4.hmm
> allseqs.hmm

Hmmbuild can also be used to concatenate new models to an existing database using the –A option. To build models out of the alignments titled seqname5.aln and seqname6.aln and automatically add them to the allseqs.hmm database, use the following commands:

HMMbuild –A allseqs.hmm seqname5.aln
HMMbuild –A allseqs.hmm seqname6.aln

...etc.

Once the models are concatenated into a database and then calibrated with hmmcalibrate, a binary index file may be created using the hmmindex program. The index file is required if you are going to use the parallel MPI or PVM versions of the program, or if you want to be able to extract the sequences individually as needed.

Using the hmmindex executable is quite straightforward, and there are no options beyond –h to obtain help. To index the allseqs.hmm file that we have just created, we will use the command

Hmmindex allseqs.hmm

Now that the database is indexed, we may extract a model from it using the hmmfetch command. Hmmfetch is a simple utility that takes a model from a database and sends it to standard output, which is usually the screen unless we direct it to a file. In order to get the seqname2 model out of our database and save it to a file named seqname2.hmm, we will use the following command:

Hmmfetch allseqs.hmm seqname2 > seqname2.hmm

In addition to extracting a model by name, we may also select it using the order in the database using the –n option. If we want to get the first model out of our database, and send it to a file called firstmodel.hmm, we use the command

Hmmfetch –n 0 allseqs.hmm > firstmodel.hmm

Notice that the numbering starts with zero in order to confuse the non-computer scientists in the crowd. Therefore the fifth model will be number 4, the ninety-second will be number 91, and so forth.

A sequence or group of sequences may be run against our shiny new database with the hmmpfam program. This is the opposite of the hmm-search program, where the HMM is the query and the fasta format sequences are the target. Still, depending on the amount of sequence data and the size of the HMM database, running the hmmPfam search can take a long time!

Hmmpfam allseqs.hmm newseqs.fa >myseqsearch.out

Starting from a group of related sequences is a useful technique, but there are other ways to proceed. If you are starting from a position of having a group of sequences that you want to search for a particular domain, then you will typically extract that domain from Pfam (or another HMM database), using hmmindex and hmmfetch, then run that model against the sequences using hmmsearch. If we want to extract the HMM for thymidine kinase, for example, we can use hmmfetch to pull it out of Pfam, providing that we have previously indexed the database with hmmindex:

Hmmindex Pfam_fs

Hmmfetch Pfam_fs tk >thymidinekinase.hmm

Please note that we have to use the value from the 'name' field from the HMM. Using thymidine or kinase as the search term will not yield any data. This is annoying when you want to fetch all the members of a class, such as all the kinases.

All griping aside, if you want to search a group of 10,000 protein sequences to see which ones might be thymidine kinases, it would be much faster to search them with hmmsearch and one model than it would be to run all 10,000 against all of Pfam. Interpreting the results is also much easier.

3.3   CREATING MULTIPLE ALIGNMENTS WITH HMMS

Both HMMsearch and HMMPfam produce an alignment between a model and a single sequence. Profile Hidden Markov Models may be used to create MSAs of a large number of sequences in a timely manner. To do this, a relatively small number of representative sequences are chosen to build an initial seed alignment. This alignment is built into an HMM using the methods we have discussed above. The larger bulk of the sequences may then be aligned with the HMM using the hmmalign program. In addition to making it possible to create an alignment with hundreds of sequences, hmmalign can show where subfamilies of proteins will form natural separations. The command for creating a MSA from a seed HMM (in this case, LEA2.hmm) and a larger group of sequences (in this example, 200LEAs.faa) is

Hmmalign –o 200LEAs.ali LEA2.hmm 200LEAs.faa

Using the –o option to save the results is necessary in this case, because otherwise the output is sent to the screen, not to a file. Watching the output of an alignment with 200 sequences scroll rapidly past your screen is not a very useful endeavor!

Using hmmalign can help you to maintain your models as the protein databases continue to grow. New hits to your models may be quickly aligned and improved without starting over from the beginning. This is how the Pfam-A database is built—using a model built with seed alignments, all other members of the family are detected with hmmsearch, then the alignments are made with hmmalign, and the final model is built out of those alignments.

3.3.1   HMMer Statistics

There is no definite theoretical method available for accurately determining E-values for gapped alignments, especially Profile-HMM alignments. HMMER uses quasi-empirical methods to estimate E-values. This simply means that a large amount of random data is generated to compare to the model, and the E-values are set based upon these data. These methods are generally fairly accurate. Generally speaking, HMMer tends to err on the conservative side.

The Pfam database annotation team has implemented another system for determining cutoff points. This system is implemented in the models themselves, rather than in the HMMer software. Two values are stored for each Pfam domain or family. These are the 'trusted cutoff' and the 'noise cutoff', and labeled TC1 and NC1, respectively. TC1 represents the lowest score for a sequence that was included in the full alignment for the family. NC1 represents the highest score for a sequence that was not included in the family. TC1 is therefore always greater than NC1.

So, how do we use this information? Most commonly, we might use the TC1 value instead of the E-value as a cutoff threshold. The logic here is that if a hit returns a score that is higher than the trusted cutoff, then that hit would likely have been included in the model by the Pfam curators.

The question often arises as to why the Pfam curators would include a sequence into the model in the first place if it does not have a low E-value! The answer is that there may be other evidence that supports the hypothesis that the sequence fits into that family. This evidence might be structural, or it might be derived from some other experimental evidence. Pfam is such a high-profile project that the curators are sure to hear about it if they do not include something that they should, or if they do include something that has been proven to be a member of a different family.

3.3.2   Sequence Scores and Domain Scores

When scoring proteins with multiple domains, we must consider another factor. Do we score the sum of *all* the domain hits, or do we treat each domain hit individually? HMMer actually does both. The total score of a sequence aligned to a particular model is the 'sequence classification score'. If this model hits in multiple places on the sequence, which generally indicates a multiple domain protein, then we may have an increased level of confidence that this sequence really does belong to that protein family, even if each individual domain hit is a weak match. The 'domain score', on the other hand, is the score for each individual domain. As you might guess, when you have a single domain protein these two scores are identical.

The E-value cutoff for an HMMPfam search uses the sequence classification score for a cutoff point. If the query sequence passes this threshold, all of the individual domains in that sequence will be reported, even if some of them have low scores. This is because some of the domains in a multidomain protein can be degraded over time, and thus match only weakly. As a result, you may see domains with crummy looking alignments and E-values that do not match your threshold cutoff point.

But how far should the search go in allowing these degraded or eroded domains into our output? For this we need another pair of scoring values for trusted cutoffs and noise cutoffs. These scores, labeled TC2 and NC2, respectively, refer to the domain cutoffs. These scores will generally be lower than TC1 and NC1, and will only be important in multidomain proteins. If a sequence finds a match to a Profile-HMM with a score that is greater than TC1 and/or NC1 then any additional individual domain scores are compared to TC2 and/or NC2.

When the sequence is compared to an HMM that represents a domain, that domain may be found several times in the sequence, as seen in figure 3.9. As

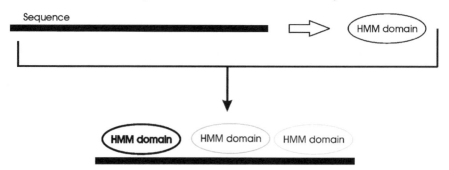

FIGURE 3.9:   A Sequence is Compared to a Domain HMM. Note That the Domain Is Found Several Times within the Sequence

long as the overall score meets the threshold, lesser domains may be included as long as they meet the TC2/NC2 criteria.

3.3.3   HMMer Format

The widely used Pfam database is generated with, and used with the HMMer package.

Compulsory fields are as follows:

NAME	Identification:	One word name for family. This is a unique identifier in that each member of a class of proteins will have a separate ID. For example, HK indicates the hydroxyethylthiazole kinase family and TK indicates a thymidine kinase.
ACC	Accession number:	Accession number in form PFxxxxx or PBxxxxxx depending on whether the model is from Pfam-A or Pfam-B.
DESC	Description:	Short description of family. The definition line gives a very brief idea of what this model represents.
LENG	Length	Indicates the length of the model, that is, the number of match states.
ALPH	Alphabet	Indicates which alphabet is used, amino acid or nucleic.
RF	Reference	Reference coordinate system for Selex format alignments.

CS	Consensus	Consensus Secondary Structure from selex format alignments.
MAP	Map annotation	If MAP flag is set to yes, allows rapid alignment to the original alignment.
GA	Gathering method:	Search threshold used in building the full alignment.
TC	Trusted Cutoff:	Lowest sequence score and domain score of match that is included in the full alignment.
NC	Noise Cutoff:	Highest sequence score and domain score of match that is not included in the full alignment.
TP	Type:	Type of family – presently Family, Domain, Motif or Repeat.
NSEQ	Sequence:	States the number of sequences that were used to build the alignment.
AM	Alignment Method	The order ls and fs hits are aligned to the model to build the full align. End of alignment.
DC	Database Comment:	Comment about database reference.
DR	Database Reference:	Reference to external database.
RC	Reference Comment:	Comment about literature reference.
RN	Reference Number:	Reference number.
RM	Reference Medline:	Eight-digit medline UI number.
RT	Reference Title:	Reference title.
RA	Reference Author:	Reference author.
RL	Reference Location:	Journal location.
PI	Previous identifier:	Record of all previous ID lines.
KW	Keywords:	Keywords.
CC	Comment:	Comments.
NE	Pfam accession:	Indicates a nested domain.
NL	Location:	Location of nested domains– sequence ID, start and end of insert.
AU	Author:	Authors of the entry.
SE	Source of seed:	The source suggesting the seed members belong to one family.

3.4 SAM

The Sequence Analysis Method (SAM) (www.cse.ucsc.edu/research/compbio/sam) has a completely separate format from HMMer. Many people prefer SAM to HMMer, and it is said to be more sensitive. In addition, SAM has the T02 program, which is an extremely powerful and useful script for finding remote homologues. It would be nice if one of you would port this script to HMMer! A conversion script is available to convert from SAM format to HMMer at the SAM website.

A study by Martin Madera and Julian Gough found that HMMer was up to three times faster than SAM on databases over 2000 sequences, while SAM was faster on smaller targets.

The SAM iterated database search system T02 has been shown to perform better than PSI-BLAST, although the computational complexity is considerably greater. As a result, SAM took over thirty times longer than the equivalent PSI-BLAST search.

Perhaps the most interesting feature of the comparison between SAM and HMMer was the observation that models made with SAM, and then converted to HMMer using the convert.pl Perl script, showed better performance than models made with HMMer. On the other hand, models made with HMMer and then converted to SAM format performed more poorly than native SAM models. This indicates that some of the sensitivity of SAM is based on its superior model-building ability.

SAM is not open-source, but a license is free to academics. Model building with SAM is made simpler with the use of a set of scripts. No model calibration program is included, or necessary. The SAM method for scoring models calculates the E-values using a scoring function that compares the difference between scores of the query sequence and its reverse, to correct for any compositional bias.

Wistrand and Sonnheimer extended the work of Madera and Gough in the studies of the differences between SAM and HMMer. This study confirmed the superior nature of the SAM model building. The conclusion of this study was that HMMer gives too much weight to sequence counts.

Structure prediction has been a major thrust and driving force behind the development of the SAM package. Members of the development team are regular participants in the Critical Assessment of Structural Prediction (CASP) competition.

3.5 PSI-BLAST, PSI-TBLASTN AND RPS-BLAST

Version 2.0 of the NCBI BLAST program was a breakthrough in many ways. For the first time, BLAST was able to handle gaps in the alignments, but for our discussion a greater improvement was the introduction of PSI-BLAST. The PSI-BLAST algorithm works as follows:

1. Begin with a single amino acid sequence.

2. Find matches using an initial protein BLAST search.

3. Align those matches.

4. Build a PSSM or PSI-BLAST profile, as it is also called.

5. Search the database again with the profile.

6. Any new hits are used to improve the profile.

7. The process is iterated until no new sequences are found. At this point, the search is considered to have reached convergence, and the iterations are stopped.

The profiles that are generated in this process may be saved for further use. Another program called RPS-BLAST (Reverse Position Specific BLAST) can be used to search this profile against any database that you like, and a collection of these profiles can provide a quick way to analyze complete genomes.

While PSI-BLAST is a fast, yet sensitive algorithm, it is not without flaws. If a domain is present that is common to a wide range of proteins, the results may be misleading, because the first search will turn up members of several protein families that will then be used to build the profiles. At times the search will not converge at all, due to the inclusion of more and more bad data. PSI-BLAST is generally run with limits set by the '-j' option to specify the maximum number of iterations to prevent the system from running endlessly.

An example of the psiblast syntax is given below:

blastpgp -d uniref90 -i test.faa -j 5

Remember that the results of a PSI-BLAST search as seen in Figure 3.10, are the alignments to the profile, not to the initial sequence. Many times the intial sequence, if it is found in the database, will not be the top hit in the final output because of this fact.

To run PSI-TBLASTN, the sequence is first run with PSI-BLAST and the '-C' option is chosen to save the position-specific scoring matrix file (that is, the profile) in the same way as one would when preparing to run RPS-BLAST. An example is given below, using the '-j' option to limit the number of iterations to a maximum of five:

blastpgp -d uniref90 -i test.faa -j 5 -C test.ckp

Then run blastall while using the '-p psitblastn' option to tell the program that you will be using a profile as a query rather than a protein sequence. All of the other parameters work the same as any other TBLASTN job, that is, when one is comparing a protein sequence to a nucleotide database:

Blastall -d est_human -i test.faa -p psitblastn -R test.ckp

In this example, the PSSM is built from data in the non-redundant version of uniprot, and then searched through the human section of DbEST.

More information on creating whole databases from the PSI-BLAST program is given in chapter 6. The related PHI-BLAST system is described below.

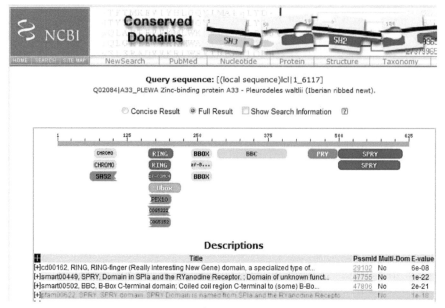

FIGURE 3.10:   RPS-BLAST Search against the CDD Database

3.6   REGULAR EXPRESSION METHODS

Computer scientists have long utilized regular expression methods for text searching with tools such as GREP. Computer languages such as Perl have regular expressions built in.

Regular expressions have many different syntax formats, and this level of non-standardization can be frustrating at times. In PROSITE format, for example, the pattern A-R-x(7,21)-Y-x-G-x(4)-D indicates that the A is followed by an R, then by a gap consisting of any letters from 7 to 21 characters long. Next follows a Y, then any character, then a G, four characters of any type, and finally a D. Ambiguity is handled within square brackets, so that [ALI] indicates that the position may have an alanine, leucine or isoleucine. While this may seem confusing for the novice user, the power of this type of search is considerable and the speed is acceptable.

Other regular expression syntax systems contain the same ideas, but the implementation is quite different. The PROSITE format inserts hyphens between each character, which is designed to increase clarity, while others do not. The ambiguity code is a lowercase x, which matches any letter in the PROSITE format, while Perl and other formats use a period for this purpose. Another difference is in the range designation—the PROSITE format as shown above uses the term (7,21) to show a range of characters but Perl syntax uses curly braces like so—7,21. PROSITE is mentioned specifically here because of its popularity, but other tools utilize different syntax systems. The Eukaryotic Linear Motif (ELM) server, for example, uses more standard posix style syntax.

The C2H2-type zinc finger domain is characterized by the following pattern of amino acid residues:

*S*tart with a C

Between 2 and 4 amino acids of any type.

Another C.

3 more amino acids of any type.

One of the following amino acids: LIVMFYWCX.

8 more amino acids of type.

One H.

Between 3 and 5 more amino acids of any type.

One final H.

This pattern, therefore, would match a sequence that would be difficult to find with a standard BLAST search:

CAASCGGPYACGGWAGYHAGWH

The PROSITE expression for this pattern is

C-x(2,4)-C-x(3)-[LIVMFYWC]-x(8)-H-x(3,5)-H

The complete PROSITE record has more information than just the pattern. The record for Colipase, which is a cofactor for pancreatic lipase, is shown below. The pattern itself is shown on the line that begins with PA, but other information may be crucial to those who wish a more in-depth assessment of the related proteins. For example, the pattern tells us that the three Y positions and the two C positions are critical, but does not tell us why. The lines beginning with CC provide this information—the former locations are involved with interfacial binding, and the latter are necessary to maintain disulfide bonds.

• PROSITE: PS00121
ID COLIPASE; PATTERN.
AC PS00121;
DT   APR-1990   (CREATED);   APR-1990   (DATA   UPDATE); JUL-1998 (INFO UPDATE).
  DE Colipase signature.
  PA Y-x(2)-Y-Y-x-C-x-C.
  NR /RELEASE=53.1,270778;
  NR /TOTAL=10(10); /POSITIVE=10(10); /UNKNOWN=0(0); /FALSE_POS=0(0);
  NR /FALSE_NEG=0; /PARTIAL=2;

CC /TAXO-RANGE=??E??; /MAX-REPEAT=1;
CC /SITE=1,interfacial_binding; /SITE=3,interfacial_binding;
CC /SITE=4,interfacial_binding; /SITE=6,disulfide; /SITE=8, disulfide;
CC /VERSION=1;
DR P02704, COLA_HORSE , T; P02705, COLB_HORSE , T; P19090, COL_CANFA , T;
DR P04118, COL_HUMAN , T; Q9CQC2, COL_MOUSE , T; P42889, COL_MYOCO , T;
DR P02703, COL_PIG , T; P42890, COL_RABIT , T; P17084, COL_RAT , T;
DR Q91XL7, COL_SPETR , T;
DR P11148, COL_CHICK , P; P11149, COL_SQUAC , P;
3D 1ETH; 1LPA; 1LPB; 1N8S; 1PCN; 1PCO;
DO PDOC00111;
//

PHI-BLAST

Pattern-Hit Initiated BLAST (PHI-BLAST) combines the matching of regular expressions with standard BLAST alignments surrounding the match. The combination incorporates the power of regular expressions with the convenience of close integration with PSI-BLAST. In fact, PSI-BLAST can be started with PHI-BLAST, with all subsequent iterations using PSI-BLAST. This is known as PHI-PSI-BLAST.

PHI-BLAST uses the same regular expression format as PROSITE. An example of the syntax is

blastpgp -i test.faa -k myseq.pat -p patseedp -d uniref90

This command will search the uniref90 database for matches to the pattern in myseq.pat and use the sequence myseq.faa for the alignments.

The choice of which regular expression system to use may not seem simple because of the many choices available. PROSITE style patterns, however, are worthy of investigation despite their variation from more standard regular expression syntax styles. This is both due to the inherent value of the PROSITE database itself, and because of the support by additional programs such as PHI-BLAST and 3motif.

3.7 MEME AND META-MEME

MEME finds motifs in groups of related DNA or protein sequences. These motifs are gapless. MEME will simply split the motif in two rather than add gaps. Meta-MEME (metameme.sdsc.edu/) takes these motifs and related sequences and builds them into Hidden Markov Models. The area between motifs is modeled imprecisely, providing a reduction in parameter space.

Therefore, Meta-MEME can provide accurate models with a smaller number of training sequences than other systems.

More information about MEME, MAST and Meta-MEME may be found in chapter 8.

3.8 WISE2

The Wise2 package (www.ebi.ac.uk/Wise2http://www.ebi.ac.uk/Wise2/) from Ewan Birney at the EBI adds some functionality such as nucleotide translation and FrameShift tolerance to the searches, but the format of the models is still HMMer style. The really interesting thing about Wise2 is that it combines gene model prediction and homology searches. The main part of Wise2 that people use is called 'GeneWise', but there are many other components that you should be aware of.

GeneWise compares genomic DNA to a protein, and can allow for frameshifts in the translation. Frameshifts are caused by genetic mutations that insert or delete nucleotides in a number that is not divisible by three. This causes the reading of codons to change so that the resulting amino acids will differ following the mutation. Stop codons may be missed, causing the protein to be longer, or new stop codons may be created, causing the protein to be longer or shorter than normal. This may result in a severe genetic disease.

The following nucleotide sequence may be translated into the following protein sequences using the frames shown below:

CATTCTTGTAAGGAAGGAG

Will translate into the following peptides:

Frame 1

H S C K E G

Frame 2

I L V R K E

Frame 3

F L Stop G R

Frame -1

L L P Y K N

Frame -2

S F L T R Met

Frame -3

P S L Q E

Inserting an additional adenine in the middle gives us CATTCTTGTAaAG-GAAGGAG. The resulting first forward frame translation changes to HSCK-GRX.

3.9   COMMERCIAL AND ALTERNATIVE HMM IMPLEMENTATIONS

SPSPfam (http://www.spsoft.com/) has been shown to provide a speedup of 3–60 times over the HMMer package. Shorter models seem to provide a greater relative speedup than longer models. This is an optimized version of hmmPfam, and no version of hmmsearch is available.

LDHMMer is another commercial package, from Logical Depth (http://www.logicaldepth.com/). This one includes both LDhmmsearch and LDhmmPfam, rewritten to provide a speedup of many times over the open-source code. This has proven to be a popular choice for those who have clusters but need higher throughput.

SledgeHMMer (sledgehmmer.sdsc.edu/) is an optimized and parallelized version of HMMer. The search is spread over several nodes through the use of the Unix locking function, splitting the sequences between the nodes in an efficient manner. The hard part is getting an even distribution of sequence data, because some sequences may be much longer than others, and therefore will take much longer. SledgeHMMer achieves a nearly optimal load balancing, even with different CPUs running at different speeds and on different operating systems. The SledgeHMMer website is unusual in that it allows the user to submit multiple sequences at once—typically webservers limit you to a single sequence to reduce the load on their servers due to the computational complexity of the searches.

ClawHMMer (http://graphics.stanford.edu/papers/clawhmmer/) uses the extremely powerful processors (GPUs) on some high-end graphics cards to accelerate the Viterbi portion of the hmmsearch code through the use of a programming language called Brook. Speeds are claimed to be 3–25 times faster than a CPU, depending on the CPU and the GPU that are being compared. ClawHMMer is an open-source project, but good graphics cards are expensive! It might be cheaper to buy a commercial solution, depending on your needs. Another point to consider is that ClawHMMer only accelerates hmmsearch, where the main bottleneck for most users is hmmPfam. Still, I hope to see more bioinformatics algorithms accelerated through the use of GPUs, as many people already have access to systems with high-end graphics systems, particularly in visualization labs. If this is the case, then ClawHM-Mer is an obvious choice.

DeCypherHMM is a hardware/software combination from TimeLogic (http://www.timelogic.com/). This is also based on the HMMer package, and the software runs on most operating systems that bioinformaticists are running—Windows, Solaris or Linux. Mac users have access through a sockets Command-Line Interface.

The accelerated code runs in custom FPGA (Field Programmable Gate Array) hardware for a speedup of hundreds to thousands of times. In addition to accelerated HMMer, ProfileSearch, ESTWise, Smith-Waterman, BLAST, FrameSearch and other algorithms are also available. The software also provides extra functionality over the open source code, and is very easy to pipeline.

Biocceleration (http://www.biocceleration.com/) has a software optimized version of HMMer. The software is part of the GenCore 6 software package, which also includes optimized versions of Smith-Waterman, ProfileSearch and GeneWise. GenCore 6 is advertised as using multithreading and highly optimized code.

CLC biosystems (http://www.clccube.com/) has a smaller FPGA system called 'the cube' that connects to any Linux system via USB 2.0 connections. Performance is said to be between that of optimized software solutions and higher-end FPGA systems.

Progeniq (http://www.progeniq.com/) also has a small, low power FPGA solution that plugs into a USB port. This device, called bioboost, runs not only accelerated HMMer, but also Smith-Waterman, ClustalW and BLAST.

At the time of this writing, there are several other groups with plans to develop hardware accelerated bioinformatics applications. Most of these involve FPGA chips, because the speedup is remarkable, and the cost and power requirements are low compared to a large server farm. The FPGA chips are available on PCI-e boards that can be added to most types of computers.

FPGAs have been known for being notoriously hard to program, but this is changing. New tools that make it easier to port programs from C to FPGA chips are getting better all the time and generic FPGA boards are available at a low cost. An open-source FPGA bioinformatics project is in the works. As a result, many popular bioinformatics applications such as HMMer may soon be available in both hardware and software versions for free, depending on your needs.

Accelerated solutions based on SIMD instruction sets are also in development by several groups. Since most modern processors have these types of options, it is reasonable to expect that this will be an area of great possibility. After all, these types of 'accelerators' are currently available by the hundreds or even thousands in large server farms. Tapping into this free resource is only prudent if your algorithm of choice supports this type of technology.

Ultimately, the optimal solution for your particular environment depends on a combination of the types of searches that you want to do, the frequency of those searches, and the hardware that you have to run them. It also depends upon your position—some vendors make their software available for free if you are an academic, which makes the price/performance ratio very hard to beat. Some people also have no time pressure, and so can afford to have a search stretch on for weeks, while others have competitors that will beat them to publication if they don't finish quickly. If you have a huge cluster that is rarely used, then there is no problem, but most of us are not so lucky!

3.10 HMMER OPTIONS

3.10.1 HMMPfam

-A <*number*> Using the –A option will limit the number of alignments to the specified number of best scoring domains. **-A0** shuts off the alignment output altogether. This option can be used to reduce the size of output files and is often used when the alignments are superfluous to the project at hand—for example, if the goal is to count the number of different classes of domains in a proteome.

-E <*number*> Set the E-value cutoff for the hit list to <*number*>, where <*number*> is a positive real number. The default is 10.0. Hits with E-values better than (less than) this threshold will be shown. More stringent threshold values (say, 0.001) will reduce the size of the output file and increase confidence in the output quality. At this threshold, a hit would be expected to have a one in a thousand chance of being found just by random chance.

On the other hand, some people will use a lenient threshold for the initial search and then use some other program to sort the output and cut off at whatever value is appropriate.

For example, let us say that a search was run with an E-value threshold of 0.00001. When the results are analyzed, some pattern emerges. The question may then arise as to whether the same pattern would appear with a more lenient threshold, say 0.01. Our poor bioinformaticist will now have to run the job again with the new threshold.

The clever researcher, on the other hand, ran the job with a threshold of 10, and saved it as tab output. When the output was brought into Excel, the threshold could be set at any number between 0 and 10 without having to rerun the job.

-T <*number*> The –T option sets the bit score cutoff threshold for the hit list to ¡ *number* >, where ¡ *number* > is any real number. The default is negative infinity because the threshold is usually controlled by the E-value and not by the bit score. Hits with bit scores higher than this threshold will be shown.

-Z ¡ *number* > This option tells HMMPfam to calculate the E-value scores as if we were using a sequence database of the specified number of sequences.

The default is arbitrarily set to 59021, which is the size of release 34 of the Swissprot database.

–acc When you are using a pipeline such as Taverna for high-throughput annotation, the output data are frequently parsed and stored in a relational database. In this case, it is better to use the HMM accession numbers rather than the names in the output reports. Using the –acc option in the search will cause the program to report HMM accessions instead of names, making them less human readable but more machine readable.

–cpu *¡ number >* On a shared machine, running large HMMPfam analyses can make many enemies. By default, HMMer utilizes all available CPUs in the server, which doesn't leave much for anyone else, now does it? There is an environment variable called NCPU that sets the number of CPUs that may be used at any one time by HMMer, but sometimes it is wise to override this value. The –cpu option allows you to do just that, giving you the appropriate amount of power for the task at hand.

-n This flag is used to indicate that the models and sequence are in nucleic acid format, not protein. This does not mean that HMMPfam will translate from nucleic to amino acid for you! To do that, you need to use another implementation such as Wise2, DeCypher or SPSPfam.

–cut_ga Some collections of HMMs are built with GA score cutoffs on every model to indicate the gathering threshold used to build the model. In other words, this was the threshold that was used for gathering the sequences to build the alignments. Adding the –cut_ga option to your search will cause HMMer to use this value for the cutoff rather than the standard E-value or bit-score cutoff.

If the models do not include the GA annotation line, the use of the –cut_ga option will have no effect.

–cut_tc Some collections of HMMs are built with TC score cutoffs on every model to indicate the trusted score threshold used to build the model. HMMbuild adds the TC annotation line if the alignment format is given in Stockholm or extended SELEX format. Adding the –cut_tc option to your search will cause HMMer to use this value for the cutoff rather than the standard E-value or bit-score cutoff.

If the models do not include the TC annotation line, then using—cut_tc will have no effect.

–cut_nc Some collections of HMMs are built with NC score cutoffs on every model to indicate the noise threshold used to build the model. HMMbuild adds the NC annotation line if the alignment format is given in Stockholm or extended SELEX format. Adding the –cut_nc option to your search will cause

HMMer to use this value for the cutoff rather than the standard E-value or bit-score cutoff.

If the models do not include the NC annotation line, then using—cut_nc will have no effect.

–domE <*number*> This option is like the –E option that is listed above, only with a focus on the individual domains rather than the entire sequence. The –domE option sets the E-value cutoff for the per-domain ranked hit list to some positive number. The default is infinity, because all domains in the sequences that passed the first threshold will be reported in the second list. This is so that the number of domains reported in the per-sequence list is the same as the number that appear in the per-domain list.

–domT <*number*> This option is like the –E option that is listed above, only with a focus on the individual domains rather than the entire sequence. The –domE option sets the E-value cutoff for the per-domain ranked hit list to some positive number.

The default is negative infinity, because all domains in the sequences that passed the first threshold will be reported in the second list. This is so that the number of domains reported in the per-sequence list is the same as the number that appear in the per-domain list.

–forward HMMer uses the Viterbi algorithm by default to calculate the per-sequence scores. The –forward option substitutes the forward algorithm for Viterbi in this step. This does not affect the Per-domain scores—they are still determined by the Viterbi algorithm. Some people think that this option should provide greater sensitivity for detecting remote homologues.

–informat ¡format> Tells the HMMer program that the input *seqname* is in a particular format. Valid format types include FASTA, GENBANK, EMBL, PIR, STOCKHOLM, SELEX, MSF, CLUSTAL, GCG, and PHYLIP.

Ordinarily, the format type is autodetected using a program called babelfish. This usually works, but not every time. To make sure that the format is specified proPerly, particularly when running large jobs that will be unattended for long periods of time, use the –informat option. This will increase the reliability slightly.

–null2 HMMer rescores each alignment with a postprocessing step by default. This takes into account that there may be possible biased composition in the HMM, the sequence, or both. Typically, the null2 correction is extremely valuable in removing false positives, particularly in the local mode. It is possible, however, that the null2 step might remove some real hits on rare occasions. The –null2 option turns off the postprocessing step, and might improve sensitivity at the cost of selectivity.

−**pvm** HMMer may be run on any cluster that has Parallel Virtual Machine (PVM) installed. The PVM must be running and the client program **hmmPfam-pvm** must be installed on all the PVM nodes. The HMM database and an associated index file *hmmfile*.gsi must also be installed on all the PVM nodes. The **hmmindex** program must be run on the hmm database to create the GSI index file.

The PVM implementation is I/O bound, so for best performance, each node should have a local copy of the HMM database rather than NFS mounting it on a shared disk. PVM support must have been compiled into HMMER for −**pvm** to have any effect.

-**h** Prints a short help file to get you started, including a summary of all options that are listed here.

3.11 SUMMARY

The HMMer package is the most popular Profile-HMM package in the bioinformatics field. Due to the open-source nature of the system, it is included in many other systems. Several auxiliary packages are part of the package, to do various types of functions, such as database searches and alignments.

SAM is the closest method to HMMer conceptually. The quality of the models is generally considered to be better than models constructed by HMMer. Conversion between the two formats is easily accomplished through the use of a simple Perl script.

PSI-BLAST is an automated method for generating profiles and using them to search databases. Sensitivity is reduced somewhat, but the speed is vastly superior to Hidden Markov Model systems. Profiles can be saved and used in PSI-TBLASTN or RPS-BLAST searches.

Regular expression searches are alternate systems to Hidden Markov Models. Several methods are available, but PROSITE is a popular database format.

MEME, MAST and Meta-MEME are motif based methods that can build Hidden Markov Models and search those models against a database. Downloads are free to academics and a website provides search capabilities with email options.

The Wise tools extend the capabilities of HMM search tools to include translation of nucleotides in all six conceptual reading frames. As a result, EST collections and even whole chromosomes may be searched against model databases. The Wise tools utilize HMMs in the HMMer format.

3.12 QUESTIONS

1. Obtain the PF00270 protein domain from the Pfam website. Use hmmemit to generate a representative sequence from this model.

2. BLAST the sequence from problem 1 against the nr database at the NCBI website, and compare the annotations with the description at the Pfam website.

3. Compare the sequence from problem 1 against the PROSITE database, and compare the results with the information from problems 1 and 2.

4. Investigate the results from the last CASP competition, and list the number of HMM-based contestants. Which system had the best performance out of this group?

5. Extract sequence number Q02084 from Swissprot and run it against the PROSITE database and report your results.

6. Take both the sequence and the PROSITE profile from problem 5 and run them both against the nr database using PHI-BLAST at the NCBI website. Report all sequence hits.

7. Were any other domains found in the NCBI search of the protein from problem 5 that were not found in the PROSITE search?

References

[1] S. Altschul, T. Madden, A. Schaffer, J. Zhang, Z. Zhang, W. Miller, and D.J. Lipman, "Gapped BLAST and PSI-BLAST: A New Generation of Protein Database Search Programs," *Nucleic Acids Res.* **25**(17), 3389–402 (1997).

[2] S. Altschul, "Amino Acid Substitution Matrices from an Information Theoretic Perspective," *J. Mol. Biol.* **219**, 555–65 (1991).

[3] S. Altschul, "A Protein Alignment Scoring System Sensitive at All Evolutionary Distances," *J. Mol. Evol.* **36**, 290–300 (1993).

[4] S. Altschul, M. Boguski, W. Gish, and J. Wootton, "Issues in Searching, Molecular Sequence Databases," *Nature Genetics.* **6**, 119–129 (1994).

[5] S. Altschul, W. Gish, W. Miller, E. Myers, and D. Lipman, "Basic Local Alignment Search Tool," *J. Mol. Biol.* **215**, 403–10 (1990).

[6] I. Korf, M. Yandell, and J. Bedell, *BLAST: An Essential Guide to the Basic Local Alignment Search Tool*, O'Reilly Media Inc., Sebastopol, CA, 2003.

[7] J. Claverie and D. States "Information Enhancement Methods for Large Scale Sequence Analysis," *Computers in Chemistry* **17**, 191–201 (1993).

[8] E. Shpaer, M. Robinson, D. Yee, J. Candlin, R. Mines, and T. Hunkapiller, "Sensitivity and Selectivity in Protein Similarity Searches: Comparison of Smith-Waterman in Hardware," *Genomics* **38**, 179–191 (1996).

[9] W. Gish and D. States, "Identification of Protein Coding Regions by Database Similarity Search," *Nature Genetics* **3**, 266–72 (1993).

[10] S. Henikoff and J. Henikoff, "Amino Acid Substitution Matrices from Protein Blocks," *Proc. Natl. Acad. Sci. USA* **89**, 10915–19 (1992).

[11] S. Karlin and S. Altschul, "Methods for Assessing the Statistical Significance of Molecular Sequence Features by Using General Scoring Schemes," *Proc. Natl. Acad. Sci. USA* **87**, 2264–68 (1990).

[12] S. Karlin and S. Altschul, "Applications and Statistics for Multiple High-Scoring Segments in Molecular Sequences," *Proc. Natl. Acad. Sci. USA* **90**, 5873–7 (1993).

[13] W. Pearson, "Searching Protein Sequence Libraries: Comparison of the Sensitivity and Selectivity of the Smith-Waterman and FASTA Algorithms," *Genomics* **11**(3), 635–50 (1991).

[14] A. Schaffer, L. Aravind, T. Madden, S. Shavirin, J. Spouge, Y. Wolf, E. Koonin, and S. Altschul, "Improving the Accuracy of PSI-BLAST Protein Database Searches with Composition-Based Statistics and Other Refinements," *Nucleic Acids Research* **29**(14), 2994–3005 (2001).

[15] D. States and W. Gish, "Combined Use of Sequence Similarity and Codon Bias for Coding Region Identification," *J. Comput. Biol.* **1**, 39-50 (1994).

[16] D. States, W. Gish, and S. F. Altschul, "Improved Sensitivity of Nucleic Acid Database Similarity Searches Using Application Specific Scoring Matrices," *Methods: A Companion to Methods in Enzymology* **3**, 66–70 (1991).

Chapter 4

HMM Databases

4.1 INTRODUCTION

Chapter 4 provides detailed information about the various types of HMM databases that are publicly available. This chapter will attempt to provide an answer to the question, "which database should I use?" and yet will emphasize that there is no single best solution for all users.

The HMM databases covered here will be discussed during the remainder of the text, and this chapter may therefore be a useful reference. The judicious use of the appropriate database can turn your sequences from a seemingly random collection of letters into a knowledge base that can greatly assist your discovery efforts. HMM databases are simply flatfile collections of Hidden Markov Models, in ASCII text format, and may be easily edited, added to, or modified although in practice they are most commonly used 'as-is'.

The Conserved Domain Database (CDD) is included here even though it contains PSSMs (Position Specific Scoring Matrices), not HMMs, because of the close conceptual relationship between the two types, and because the CDD is partially built from other popular HMM databases.

4.2 THE MANY FLAVORS OF PFAM

Pfam is a joint project from the Sanger Centre, Washington University, Karolinska Institute, and INRA, although the main location is at the Sanger Centre. Pfam contains models of thousands of protein clans, families, domains and motifs. In addition, there are two versions of Pfam, built with two different methodologies. Pfam-A is hand curated from custom multiple alignments. Pfam-B is generated automatically from ProDom. Earlier versions of ProDom were created using the Domainer algorithm from the SP/TremBL database, but more recent editions have replaced Domainer with Psi-BLAST. Domain families that have been found with PSI-BLAST are then clustered with Mkdom2 and aligned with ProDomAlign, which is well suited to aligning large numbers of sequences.

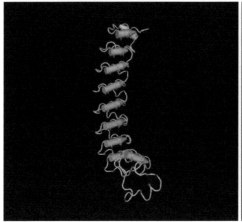
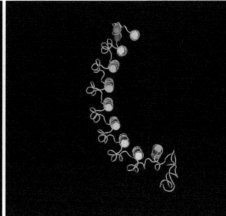

FIGURE 4.1:   The LRV Domain—Three Views

Approximately 75% of sequences will have at least one match to Pfam-A. Another 19% have a match to Pfam-B. This does not mean that those matches tell everything about that protein sequence, however. There are hundreds of Pfam entries that include the term 'unknown', as in 'Domain of Unknown Function' (frequently shortened to DUF), or 'Protein of Unknown Function'. Perhaps more commonly, some information is known about the protein, but it has not been fully elucidated. For example, Pfam entry PF01816 is a Leucine-Rich Repeat Variant, or LRV. Its structure has been known for years, and the model from the Protein Data Bank (PDB) is shown in figure 4.1. These repeat regions may be found in a number of different types of proteins, with functions as diverse as cell adhesion, DNA repair, and hormone reception. It appears that this repeat region is involved with protein-protein interactions, and the important point to make here is that the simple identification of this entry in a sequence of interest cannot indicate the function of the overall protein.

It is also important to note that there are two versions of both the Pfam-A and Pfam-B databases. Pfam-ls is designed for global alignments, as defined earlier, and Pfam-fs is optimized for local alignments so that matches may include only part of the model. Both the –fs and the –ls versions are local with respect to the sequence. This allows the models to be used with assembled data, such as chromosomes or genomes.

Pfam benefits from the fact that it is extremely widely used. People with expert knowledge of various protein domains will offer their opinions on the appropriate Pfam records, making the collection more useful for all of us. For example, an expert in a particular type of tyrosine kinase might find that a sequence of interest has been left out of the alignment for the appropriate tyrosine kinase model. That person may then contact the Pfam curators to have the sequence included, and the model will be improved in the next release of Pfam.

Pfam models represent families, domains, repeats and motifs. The type of model may be found in the annotation on the description line, as shown below.

DESC Protein kinase C terminal domain

DESC FNIP Repeat

Closely related to Pfam is iPfam, which describes domain-domain interactions that are observed in PDB entries. PDB stands for 'Protein DataBank', and is the repository for protein structures that are derived from crystallographic and NMR studies. Most PDB domains are defined in Pfam entries. The algorithm for determining domain-domain interactions starts with mapping the Pfam domains to PDB structures, then identifying bonds between the domains.

While many people prefer to search Pfam locally using the HMMer Package, the Pfam website (www.sanger.ac.uk/Software/Pfam) has many links, features, and additional tools that are extremely useful. A good rule of thumb is to search Pfam with a large number of sequences locally, and then go to the website to identify supplemental information for sequences of interest. If you have already identified a few sequences of particular interest, then go directly to the Pfam website—the additional features found on the website as opposed to running them locally will be well worth it.

4.3 SMART

The Simple Modular Architecture Research Tool (SMART) provides a database of models developed with an emphasis on mobile eukaryotic domains.

SMART is available at smart.embl-heidelberg.de, and is annotated with respect to phyletic distributions, functional class, tertiary structures and functionally important residues. Use SMART for signaling domains or extracellular domains. Note that SMART may be used in either Normal or Genomic mode. The difference between these two is the underlying databases. The Normal mode uses Uniprot data along with stable proteomes from Ensembl. This can lead to redundancy, although the curators do remove identical sequences, different protein sequences and sequences of protein fragments can all belong to the same gene. This problem can throw off the counts and statistics of domain architectures. The genomic mode avoids this problem by only utilizing data from complete proteomes.

SMART can now be queried using Gene Ontology (GO) terms, making it particularly useful to those looking for specific categories of domains. The entries in SMART may be searched for disordered regions as predicted by DisEMBL's hotloops and REM465 programs. The website provides easy analysis of domain interactions and provides the network context for hundreds of thousands of proteins based on the STRING database.

Predictions of catalytic activity are given for several dozen domains based on the presence of key residues. Domains that are similar to certain enzymes have been found to have no catalytic activity when these key residues are not conserved. Including key amino acid residues in SMART allows a more accurate analysis of putative enzyme homologs.

The order of the domains in a protein sequence is referred to as the domain architecture. SMART takes all the protein sequences with the same domain architecture and maps them onto the NCBI taxonomy. The last common ancestor of all the organisms that contains that architecture then predicts the origin of the domain architecture. We may then make inferences that this architecture may be found in genomes that are not yet sequenced.

The authors of SMART have used it as a tool for discovery, and some of the output is available for download. For example, the entire Yeast genome has been checked for signaling domains, and the output is retrievable.

SMART is a much smaller collection than Pfam, and is therefore not an 'all-purpose' system for analysis. The domain architecture features, GO links, annotations of regions of Intrinsic Disorder and catalytic activity prediction make it extremely useful despite the lesser size.

Figure 4.2 shows a representation of an enteropeptidase precursor as visualised by SMART. Figure 4.3 demonstrates a number of domain assignments as shown by SMART.

FIGURE 4.2: SMART Representation of Human Enteropeptidase Precursor

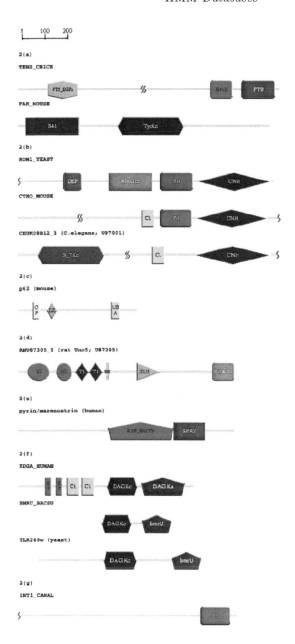

FIGURE 4.3: Examples of SMART Domain Assignments

4.4   TIGRFAM

The Institute for Genomic Research (TIGR) has been merged to form part of the J. Craig Venter Institutes. They have produced the TIGRFAMs database of Hidden Markov Models. These models are manually curated, and are organized by functional role. TIGRFAMs are partially based on the concept of equivalogs, a set of homologous proteins that are thought to have a conserved function since their last common ancestor. Similarly, equivalog domains are domains of conserved function. The TIGRFAMs database also contains full-length protein models at the levels of superfamilies and subfamilies. The term 'superfamily' here represents the protein homology over essentially the entire length of the protein. Members of a superfamily do not necessarily share the same function. Several clades may be found within a superfamily, with each one represented by a subfamily HMM.

Version 4.1 of TIGRFAMs contains 2453 models, with an emphasis on identifying genes in microbial data. These models are designed to be complementary to Pfam, so these two databases should be run together as a set. TIGRFAMs are part of the Comprehensive Microbial Resource (CMR), and help to categorize genes in terms of roles and subroles. For example, under the role of fatty acid and phospholipid metabolism the CMR lists the subroles of biosynthesis, degradation or other. These roles and subroles are listed for several hundred genomes, so check the list of completed analyses before running any searches—your project may already be finished.

A sequence from the rat was compared to both Pfam and TIGRFAM, with the results shown graphically in figure 4.4. While the upper portion of the figure shows that the sequence is most likely a pyruvate carboxylase, the lower portion breaks this analysis down into constituent domains. The 'best' representation depends upon your specific needs. Some people study thousands of protein sequences at a time, and for these types of projects a single overall descriptor is probably most useful. At other times, you may wish to study a single protein in great depth, perhaps to study interaction binding sites or to find out in some other way how it works, not simply what it does. For this type of study, looking at the domain structures is more useful than a single designator.

4.5   SUPERFAMILY

The SUPERFAMILY database is maintained by Martin Madera at the MRC Laboratory of Molecular Biology and Julian Gough, formerly at the MRC and now at the University of Bristol. A web server providing access

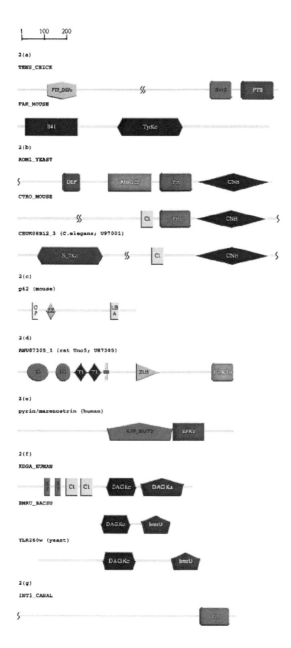

FIGURE 4.4: TIGRFAM and Pfam Alignments for Pyruvate Carboxylase. The thin line represents the sequence. The bars represent hit regions.

Complete Genomes

Click on a genome name	Dom	No. of sequences	No. with assignment	% with assign.	% total sequence coverage	No. of domains assigned	No. of supfam.s	Average Supfam size	% produced by duplication	Average sequence length	Average length matched	No. of domain pairs
Homo sapiens 22.34d	E	34091	22375	66	40	53081	953	55.7	98	496	302	1382
Pan troglodytes 22.1	E	38822	21583	56	40	42894	925	46.4	98	343	245	1245
Mus musculus 22.32b	E	32281	21104	65	43	46177	963	48	98	429	285	1333
Rattus norvegicus 22.3b	E	28545	19624	69	45	44260	937	47.2	98	441	286	1327
Gallus gallus 22.1	E	28416	18169	64	40	38050	867	43.9	98	503	315	1243
Xenopus tropicalis 2.0	E	28382	19398	68	40	35822	958	37.4	97	503	294	1991
Fugu rubripes 22.2c	E	33003	22692	69	42	52584	882	59.6	98	550	339	1320

FIGURE 4.5:   Genomic Analysis with SUPERFAMILY

to the SUPERFAMILY database is at http://www.suPfam.org/. The models may also be downloaded for non-commercial use. A license is required, but it is easy and free to apply, and there is no wait for approval.

SUPERFAMILY is built from the Structural Classification of Proteins (S-COP) database, which is in turn built from the Protein Database (PDB). At the time of this writing, there are 1447 SCOP superfamilies, each represented by a group of models, for a total of over 8500 models. In addition, the authors provide a useful spreadsheet that compares the SUPERFAMILY designations to Pfam, InterProScan and Gene Ontology.

The authors of SUPERFAMILY have run a wide range of genome data through the system and present the results sorted by complete genomes, s-trains and others. This table, part of which is shown in figure 4.5, may be sorted by any one of a number of different parameters, depending on interest.

One of the interesting aspects of the SUPERFAMILY models is that they are available in SAM, HMMer and PSI-BLAST formats, so that users may choose their favorite application. Rather than change the pipeline to match the database, one may choose the database format to match the analysis pipeline.

When comparing two or more bioinformatics datasets, there is a problem that frequently crops up in that there may be overlap between the two. What is most desired is a table that provides the mapping of the identifiers of the first database with the identifiers of the second. The authors of SUPERFAMILY provide a table with mappings from their database to InterPro, Pfam and Gene Ontology (GO). An example portion of this table is shown in figure 4.6.

The first column of the map table gives the SUPERFAMILY identifier. The map column gives the nature of the mapping. If the map column shows the character A, then there is a one to one mapping between the Superfamily id and GO. If the column contains a B then two or more GO members map to a single SUPERFAMILY entry, and C is the opposite of B in that two or more SUPERFAMILY members map to a single GO entry. A D indicates that there are multiple mappings in both directions and E indicates that there is no mapping at all.

The Gene Ontology database places genes into categories of Function, Process and Cellular Location. These categories are listed as F, P and C, respectively, in the map table. Note that it is frequently not possible to assign

Map

A: denotes a 1 to 1 mapping
B: denotes multiple PFAM families mapping to a single SCOP superfamily
C: denotes a PFAM family mapping to multiple SCOP superfamilies
D: denotes multiple mappings in both directions
E: denotes no mapping

Superfamily map	InterPro-entry	PFAM-family	GO	GO-identifier	GO description	Superfamily description
52440	D	IPR000291	PF01820	F	GO:0008716	D-alanine-D-alanine ligase	Biotin carboxylase N-terminal domain-like
52440	D	IPR000291	PF01820	P	GO:0009252	peptidoglycan biosynthesis	Biotin carboxylase N-terminal domain-like
56059	D	IPR000291	PF01820	F	GO:0008716	D-alanine-D-alanine ligase	Glutathione synthetase ATP-binding domain-like
56059	D	IPR000291	PF01820	P	GO:0009252	peptidoglycan biosynthesis	Glutathione synthetase ATP-binding domain-like
47686	A	IPR000020	PF01821	C	GO:0005576	extracellular	Anaphylotoxins (complement system)

FIGURE 4.6:   SUPERFAMILY Mapping to InterPro, Pfam and GO

all three characters to a given gene. In figure 4.6, note that SUPERFAMILY members 52440 and 56059 map to a single InterPro entry (IPR000291), a single Pfam entry (PF01820) and several GO entries. The mapping is listed as D, indicating the many-to-many relationship. Entry 52440 maps to a ligase for the function, while the process is listed as peptidoglycan biosynthesis. The cellular location is not identified for this entry.

4.6   PANTHER

The PANTHER (Protein ANalysis THrough Evolutionary Relationships) database is a resource that has been developed by researchers at Applied Biosystems and released for public use. PANTHER categorizes proteins as family and subfamily. Families are evolutionarily related proteins; subfamilies are related proteins with the same function. PANTHER is also annotated as to molecular function, biological process and pathway. These are defined as follows:

Molecular function: The function of the protein by itself or with directly interacting proteins at a biochemical level, e.g. a protein kinase.

Biological process: The biological process specifies the function of the protein in the context of a larger network of proteins that interact to accomplish a process at the level of the cell or organism, such as mitosis.

Pathway: Similar to biological process, but a pathway also explicitly specifies the relationships between the interacting molecules. Biological pathways may be metabolic pathways, developmental pathways, genetic regulatory networks and signal-transduction pathways.

The total size of the PANTHER database is impressive! Version 6.1 contains 5547 protein families. Within these families are 24582 functionally distinct protein subfamilies. The subfamilies are represented within the family by a phylogenetic tree that defines the relationships between subfamilies. Families and subfamilies are both represented through the use of HMMs and are associated with ontology terms.

PANTHER has a special ontology system called PANTHER/X. This is a controlled vocabulary of terms relating to molecular function and biological process. PANTHER/X is related to the Gene Ontology (GO) system, but is highly simplified and abbreviated. The creators of PANTHER/X intended to produce a system that would be more amenable to high-throughput analysis pipelines than the more common GO terms. Mappings are available to convert from PANTHER/X to GO, and vice-versa.

PANTHER/X ontology types, as mentioned above, involve Biological Process, Molecular Function and Pathway, as opposed to the GO categories of Biological Process, Molecular Function and Cellular component. So, the information is not directly transferable.

The PANTHER website (www.pantherdb.org) uses an interactive Java applet called prowler that may be used to search the PANTHER ontologies, select specific terms, and download data associated with the various terms, such as individual genes, subfamilies or whole families of data. Being able to search by pathway is a remarkably useful feature—generating a list of candidate gene targets for a particular disease related pathway is a simple task using PANTHER.

Pathways in PANTHER may be curated by members of the community. To avoid redundancies, the pathway schedule may be viewed, so that you may see what pathways are completed or in process. You may contribute your own pathway diagrams by downloading the CellDesigner program, which is freely available for Windows, Macintosh and Linux. If you want to add more functionality, the Systems Biology Workbench works with CellDesigner to add simulation and analysis features.

Gene expression values may be evaluated using PANTHER by mapping a list of genes from a microarray experiment to a PANTHER ontology. The gene expression values may then be overlaid onto a pathway diagram with genes colored differently according to the gene clusters that they are associated with.

Gene expression values may also be uploaded to the PANTHER website for a Mann-Whitney U test. This method is able to find correlations that are missed with simple cutoffs, for example, by only selecting genes that have changes in expression exceeding 2 or 3 fold. This test is based on the probability that a particular set of genes would have random values equal to those found by the test.

The PANTHER database is generated through the clustering of the UniProt database using a BLAST-based similarity score. A dendrogram is generated that is then used to define the protein families.

Due to the sheer size of the PANTHER database, searches could be slow. The PANTHER administrators have developed a clever method of speeding up the searches. First, sequences are BLASTed using relaxed criteria against consensus sequences that represent the models. Then, searches are made only against models represented by those hits. This method drastically reduces the total search time.

Of course, total database size should not be considered a measure of overall quality or usefulness. In order to test whether PANTHER could be used to annotate proteins that were otherwise difficult to match, 3451 protein sequences from *Arabidopsis thaliana* were tested against the PANTHER database. These sequences had no matches to Pfam, SUPERFAMILY, S-MART or TIGRfam. One hundred and sixty of these sequences had significant matches to PANTHER, demonstrating that PANTHER may be a useful tool for discovering the function of heretofore 'unknown' genes.

PANTHER is incrementally being added to the InterPro suite of databases, and there is a mapping file to compare the two systems.

4.7 PRED-GPCR

The PRED-GPCR system from Papasaikas et al. at the University of Athens is designed to predict G-Protein Coupled Receptors and the particular type of GPCR that the sequence resembles. The categorization scheme is based upon the TiPs Pharmacological classification. As a result, 67 GPCR families are represented by 265 HMMs. The families are further grouped into five classes—Rhodopsin-like, secretin-like, metabotropic glutamate/pheromone, cyclic nucleotide receptor and frizzled/smoothend family.

The HMMs are regularly updated, and are available on the website at http://bioinformatics.biol.uoa.gr/PRED-GPCR/. See an example screenshot in figure 4.7. The searching includes an optional low complexity pre-processing step using the CAST algorithm. The entire system is regenerated annually.

Map

A. denotes a 1 to 1 mapping
B. denotes multiple PFAM families mapping to a single SCOP superfamily
C. denotes a PFAM family mapping to multiple SCOP superfamilies
D. denotes multiple mappings in both directions
E. denotes no mapping

Superfamily map	InterPro-entry	PFAM-family	GO	GO-identifier	GO description	Superfamily description
52440	D	IPR000291	PF01820	F	GO:0008716	D-alanine-D-alanine ligase	Biotin carboxylase N-terminal domain-like
52440	D	IPR000291	PF01820	P	GO:0009252	peptidoglycan biosynthesis	Biotin carboxylase N-terminal domain-like
56059	D	IPR000291	PF01820	F	GO:0008716	D-alanine-D-alanine ligase	Glutathione synthetase ATP-binding domain-like
56059	D	IPR000291	PF01820	P	GO:0009252	peptidoglycan biosynthesis	Glutathione synthetase ATP-binding domain-like
47686	A	IPR000020	PF01821	C	GO:0005576	extracellular	Anaphylotoxins (complement system)

FIGURE 4.7: The PRED-GPCR Webserver System

4.8   THE CONSERVED DOMAIN DATABASE, CDD

The Conserved Domain Database (CDD) from the NCBI is not an HMM database, but rather a database of Position Specific Scoring Models, or PSSM-s. These PSSM profiles are closely related to HMMs but are searched either with Impala or RPS-BLAST (Reverse Position Specific BLAST). RPS-BLAST is much faster than Impala, and so is much more commonly used.

Using the CDD with RPS-BLAST is a very fast way to analyze signature databases. The CDD curators take data from COG, KOG, LOAD and Pfam, eliminate the redundancies and add some models of their own, to come up with over 17000 models (as of version 2.11). This is a very fast way to search several major databases in a much faster manner than any of the HMM programs, although some loss of sensitivity may be experienced.

The CDD is searched as a default when running protein BLAST at the NCBI website. When you run a protein BLAST search, the first thing you see is the domain information that is predicted for your sequence, as in figure 4.8. This information comes from the CDD, and the prediction is made with RPS-BLAST. Many people will click right past this output to go to the 'real' BLAST output, which is frequently less informative than the domain information that they just skipped! Links are provided between the CDD and 3D data where possible.

NCBI runs CDD searches on their data on a regular basis. The pre-computed results are then available very quickly when a search is made. A related tool called CDART (Conserved Domain Architecture Retrieval Tool) is used to search for proteins with similar domain architectures, which are the sequential order of protein domains. Since CDART is searching through domain profiles rather than through sequences, the service is extremely fast. Since the domains are usually well annotated, the matches are quite informative, as shown in figure 4.9.

Since the CDD is based on PSSMs, not HMMs, one might wonder if they are somewhat lesser in quality. In particular, the RPS-BLAST program, being word-hit based, might be less sensitive than HMMPfam. One might expect this to be the case, particularly on shorter models, due to the heuristic nature of the RPS-BLAST search, but to the best of my knowledge there has never been a study done to investigate this. The obvious choice of database for a test like this would be SUPERFAMILY, since it is available in PSSM, HMMer and SAM formats.

FIGURE 4.8:   Conserved Domain Output

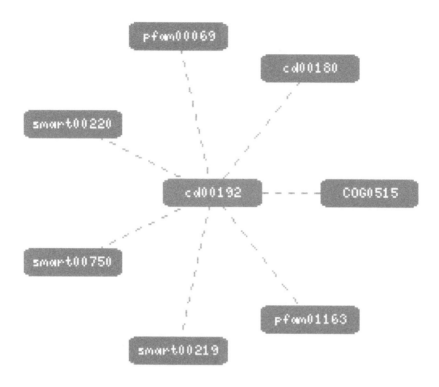

FIGURE 4.9: Domains Related to Tyrosine Kinase C

4.9 COG

The Clusters of Orthologous Proteins (COG) project at the NCBI is a phylogenetic classification of proteins encoded in complete genomes. The COGs were determined by comparing protein sequences encoded in forty-three complete genomes, representing thirty major phylogenetic lineages. Each COG consists of individual proteins or groups of paralogs from at least three lineages and thus is believed to represent an ancient conserved domain.

The COG database builds about 150,000 proteins into about 5,000 clusters. These data are available from the NCBI at http://www.ncbi.nlm.nih.gov/COG in sequence format and may be found in RPS-BLAST format and downloaded at http://www.ncbi.nlm.nih.gov/Structure/cdd/wrpsb.cgi.

I have built these data into a Hidden Markov Model database, and this may be obtained from sourceforge.net.

4.9.1   KOGs and TWOGs

Two more recent projects along these lines are the KOGs and the TWOGs. KOGs are built from purely eukaryotic data, and involve a number of additional steps other than the standard COG process. While it may be correct to consider the KOG database as a Eukaryotic COG, careful masking of repetitive domains must be carried out, or overclustering can occur. The TWOGs, on the other hand, are clusters from two lineages, and may be considered potential cogs. With additional data, it is quite likely that the TWOGs will eventually become COGs.

4.10   THE TLFAM DATABASE

Hidden Markov Models are usually designed with a broad range of training data in order to maximize the ability to find remote homologues. This allows the more general model to find matches in the widest range of data. The TLFAMs data sets are designed to turn this idea around completely. The idea behind the TLFAMs system is that since HMMs reflect the training data, specific training sets provide better results for that model when used on the same type of organism. So, this series uses Archaeal data as a training set to build models that will be used to study new archaeons, fungal data to study fungi, etc.

Not all proteins are well represented in all types of organisms—since mammals are the subject of a great deal of sequencing studies, they are better represented in the databases than archaeons, for example. This is changing rapidly, however, now that the rate of data generation is so high. Still, some taxa datasets may only have one or a very few representatives of a particular protein family or domain. Rather than lower the stringency of the gathering step and allow more questionable data into the training set, it was decided that these data would be simply dropped from the TLFAM series. This is because the TLFAMs (like the TIGRFAMS) are designed to be used in conjunction with Pfam, not as a replacement for Pfam.

An earlier version of the TLFAM prokaryote database was used to study a bacterial proteome that was newly released, and was not included in the training dataset. TLFAM showed higher scores and better alignments than Pfam for these data, but Pfam had more total hits, demonstrating the need to use Pfam along with TLFAM, instead of running TLFAM in stand-alone mode. *P. falciparum*, a eukaryote, was used as a negative control for the prokaryotic database. In this case, Pfam showed better scores and alignments, as predicted.

More recent members of the TLFAM series of databases have been generated with specific projects in mind. For those working on the specific projects for which they are intended, these are useful tools, but for the average researcher they are probably not useful. These include the following:

- The Epsilon family database. Epsilon proteobacteria are endosymbiotic with a number of other organisms, particularly gastropods at the openings of hydrothermal vents.

- The CPTfam database consists of a series of Cis-prenyl transferase models and is used to assist in the characterization of rubber biosynthesis genes.

- The Allerfam database was also designed to assist the rubber biosynthesis project. In this case, the twelve protein types known to be involved in latex allergies were built into Hidden Markov Models in order to provide a resource for the identification of potential latex allergens in novel latex sources such as Guayule.

- HydroHammer was originally designed to characterize Late Embryogenesis Abundant (LEA) proteins, but was later expanded to include all hydrophilins.

The TLFAM series is available at sourceforge.net.

4.11 KINFAM

The protein kinase family contains hundreds of members in humans, of which many are likely to be druggable targets for therapeutic purposes. The KinFam database models represent various classes of kinases. Searches against the KinFam database will assign the candidate sequence to the class and group of kinase based upon the Hanks classification scheme. This provides a more detailed look at the protein than other tools that may use a broader classification scheme.

As a test of the KinFam database, a sequence from a fibroblast GF receptor was submitted to a Pfam search. The top hit was correctly identified as a Protein Tyrosine Kinase, and the second entry was for a 'protein kinase domain.' This is correct—a Fibroblast Growth Factor Receptor is indeed a tyrosine kinase, but these results do not indicate what type of tyrosine kinase it might be classified under.

When the search was run with KinFam, as shown in figure 4.10, the top hit was to a Fibroblast GF receptor. The second match was to a different type

RANK SCORE QF TARGET—ACCESSION E_VALUE
DESCRIPTION
852.93 1 KinFam——ptkgrp15 9.3e-256 Fibroblast GF recep
479.14 1 KinFam——ptkgrp14 3.1e-143 Platelet derived GF
423.33 1 KinFam——ptkother 1.9e-126 Other membrane-span

FIGURE 4.10: Sample Output from the KinFam Database

of growth factor receptor. While the more general Pfam search was correct, the KinFam search provided a better picture of the protein function.

4.12   PRIAM AND METASHARK

Now that many genomes have been completely sequenced, the metabolic pathways may be predicted by identifying the enzyme coding genes. Having a method to automatically assess the class of enzyme rather than a simple match with another set of sequences would provide a more systemic view of the protein.

The ENZYME database from ExPASy (Expert Protein Analysis System) provides nomenclature for all enzymes for which an Enzyme Commission number has been assigned. PRIAM (PRofils pour l'Identification Automatique du Mtabolisme) is a system for automated detection of enzymes in a fully sequenced genome using the ENZYME classification system. The sequence data from the ENZYME database are clustered with the MKDOM2 program, and then profiles are built for each class of enzyme using PSI-BLAST.

The PRIAM profiles may be used with RPS-BLAST to identify enzyme sequences in newly sequenced genomes, and to classify those enzymes based on the ENZYME database methodology. PRIAM may be used to map data onto KEGG (Kyoto Encyclopedia of Genes and Genomes) metabolic charts to enable easy interpretation of the pathways that involve the protein family.

The MetaShark system (Metabolic Search and Reconstruction Kit) takes this idea a step further. MetaShark starts with the sequences used in PRIAM, aligns them with a MSA tool called MUSCLE, and builds them into a series of HMMs using the HMMer package. MetaShark may be used through any standard browser, and a small Java applet called SharkView is used to visualize the resulting networks, enzymes, reactions and compounds. The SharkHunt package may be used locally to predict metabolic networks automatically from DNA sequence, using the Wise2 tools and the KEGG database.

The advantage that MetaSHARK has over PRIAM is that you do not have to run a gene finder and translation program before running the scan. The disadvantage is that MetaSHARK is much slower than PRIAM, with one test showing approximately 100 times difference between the two.

The HMMs from MetaSHARK may be concatenated into a standalone H-MM database, but this is not as useful as running the entire system, as the annotation provides only the Enzyme Commission (EC) number. The EC number may then be searched for information on name, enzyme class, reaction, product, cofactors, pathways, orthology, genes, structures and references. For a single enzyme sequence, this is a wealth of information! Unfortunately, for a large-scale analysis, the process becomes too tedious. This is only really useful if you build a tool with links to PDB, BRENDA, KEGG pathways or some other system.

4.13   NODE

Protein function is typically thought of as relying upon its structure. Substitutions in the primary amino acid sequence that do not significantly alter that structure are more commonly seen because the function is not altered. Amino acid substitutions that alter the structure tend to change the function of the protein, causing a disruption in the cell.

Disordered proteins can take multiple conformations, which allows a single protein to undertake a number of different tasks. These adaptable proteins will sometimes take a fixed structure only when interacting with other proteins. As a result, natively disordered proteins are less studied than those with a fixed crystalline structure, but may be equally important in a number of cellular functions.

It is important to note that proteins are not typically entirely structured or entirely disordered, but rather contain regions of both. The Protein DataBank (PDB) refers to these regions with 'Remark 465' designations.

Several programs have evolved to predict areas of native disorder from primary amino acid sequence. The majority of these algorithms utilize neural networks to make assignments. The Nevada Order/Disorder Evaluator (N-ODE) uses HMMs to make predictions of disordered regions. The increased speed of the HMMs compared to the neural net method allows the rapid searching of an entire proteome in a reasonable amount of time, particularly if optimized software or an accelerator is used.

4.14   FPFAM

Data from thirty fungal genomes were used to make a more specific library of HMMs that has been named FPfam (for Fungal Pfam). A fungal Maize pathogen has been included, as has the white rot fungus, *Phanerochaete chrysosporium*. Two non-fungal plant pathogens were also included in the source dataset.

The predicted proteins of these genomes were clustered with the mkdom program. This provides an automated method that is similar in construction to the Pfam-B database.

FPfam has been tested on five newly sequenced fungal genomes, and has demonstrated improved coverage, higher bitscores and more domains per sequence than Pfam for the same domains. On the other hand, the greater specificity of the models means that some sequences are missed by FPfam that are found by Pfam. This is consistent with earlier tests with the TLFAM databases, and indicates that both FPfam and Pfam should be used to annotate new fungal sequence data. The authors of FPfam have shown that while their library provides superior coverage compared to Pfam, the combination of the two provides better coverage than either one used alone.

The authors of FPfam have shown instances of hits that their database can detect that Pfam will miss, for example, LICD proteins, members of the laminin family, and some Ribosomal_S6 proteins.

FPfam has also been shown to detect more multiple domain proteins per sequence than Pfam, demonstrating that it can find more domains per sequence as well as simply finding more single domain proteins. This effect was demonstrated at three different cutoff values to ensure that the results were not restricted to a single E-value.

4.15   KINASEPHOS

Several essential cellular processes such as cell signaling, differentiation, metabolism and membrane transport are affected by the phosphorylation of proteins. These proteins are phosphorylated by a variety of protein kinases at a serine, threonine or tyrosine residue, but not all of these residues correspond to phosphorylation sites. Phosphorylation is the most common type of post-translational modification, and the most important method of cellular regulation. The enzymes that are involved in this process need to be highly specific, that is, they need to work with one site, but not all.

KinasePhos is a method to computationally predict phosphorylation sites in a set of protein sequences using Hidden Markov Models. The HMMs are constructed from all sequences surrounding known phosphorylation sites. Swissprot entries referring to phosphorylation sites without experimental verification are not used in this training set. Instead, those sequences that are annotated 'by similarity' or 'probable' are used as a test set, not a training set. Similarly, serine, threonine and tyrosine sites that are not phosphorylation sites are used as a 'negative' set.

Larger families were split into subgroups, and models were validated by both leave-one-out cross validation and k-fold cross validation. Once the models are fully evaluated, the models with the best results are chosen.

4.16 SUMMARY

The number of HMM databases has increased rapidly in the past few years, and the growth rate is not likely to slow. Despite predictions that the total number of domains would reach a plateau at some point, recent sequencing projects have demonstrated that this point has clearly not yet been reached.

While the databases described in this chapter represent a tremendous amount of information, new genomes still have a significant number of genes that are not identified by any of these tools. Therefore it is clear that a vast amount of work is yet to be done in this area. Furthermore, a match to one or more of the models in these collections does not mean that the gene is understood. Hundreds of models are annotated as 'Domains of Unknown Function' or 'Proteins of Unknown Function.' Furthermore, hundreds of additional protein domains have very little information. These regions will need to be characterized in the laboratory, thus improving the annotations in many genomes at once.

While Pfam is an excellent resource, and it is certainly the most widely used of all the tools listed here, the astute researcher will make an intelligent selection of which database will provide the best results for the specific goals of the project at hand. The eager computational biologist will occasionally be tempted to utilize a 'kitchen sink' approach which would involve including all databases for every search. This will not be the most useful method, particularly if a large amount of data is to be analyzed. The searches will likely be excessively slow, even on large compute clusters.

A better method would be to choose carefully which target to use out of the choices presented here. If there is some doubt, a subset of the sequence data to be searched may be run against the various choices, to provide some feedback about the utility of each database. This will provide feedback about which direction the rest of the project should take.

These databases were developed for specific purposes and with specific methodologies in mind. You may well find that none of them are exactly correct for your purposes. For this reason, chapter 6 is designed to help you build your own databases for any particular purpose that you may have in mind.

4.17 QUESTIONS

1. Which databases are included in the InterPro system?

2. Pfam provides some of the domains for CDD. What other components make up the CDD database?

3. Look up InterPro entry IPR000239 and discuss which databases (other than InterPro) you might use to search for this type of protein in a set of sequences.

4. What is the relationship between EC numbers and KEGG pathways, and how might you assign them to a group of protein sequences?

5. Discuss some of the advantages and disadvantages of the Pfam database compared to the Conserved Domain Database.

6. Your instructor will provide you with a set of sequences. Search them against various web resources and compare and contrast the output from the different tools.

7. Does the PANTHER method of breaking families into subfamilies violate the rule of Entropy Maximization given in chapter 2? Why or why not?

8. Discuss the differences between Pfam-A, Pfam-B and Pfam-C, using the Pfam website to go beyond the information presented here.

References

[1] J. Gouzy, F. Corpet, and D. Kahn, "Whole Genome Protein Domain Analysis Using a New Method for Domain Clustering," *Computers and Chemistry* **23**, 333–340 (1999).

[2] I. Alam, S. Hubbard, S. Oliver, and M. Rattray, "A Kingdom-Specific Protein Domain HMM Library for Improved Annotation of Fungal Genomes," *BMC Genomics* **8**, 97 (2007).

[3] D. Haft, B. Loftus, D. Richardson, F. Yang, J. Eisen, I. Paulsen, and O. White, "TIGRFAMs: a Protein Family Resource for the Functional Identification of Proteins," *Nucleic Acids Res.* **29**(1), 41–43 (2001).

[4] A. Bateman, E. Birney, R. Durbin, S. Eddy, K. Howe, and E. Sonnhammer, "The Pfam Protein Families Database," *Nucleic Acids Res.* **28**(1), 263–266 (2000).

[5] A. Bateman, E. Birney, L. Cerruti, R. Durbin, L. Etwiller, S. Eddy, S. Griffiths-Jones, K. Howe, M. Marshall, and E. Sonnhammer, "The Pfam Protein Families Database," *Nucleic Acids Res.* **30**(1), 276–280 (2002).

[6] S. Pandit, D. Gosar, S. Abhiman, S. Sujatha, S. Dixit, N. Mhatre, R. Sowdhamini, and N. Srinivasan, "SUPfam—A Database of Potential Protein Superfamily Relationships Derived by Comparing Sequence-Based and Structure-Based Families: Implications for Structural Genomics and Function Annotation in Genomes," *Nucleic Acids Res.* **30**(1), 289–293 (2002).

[7] M. Madera and J. Gough, "A Comparison of Profile Hidden Markov Model Procedures for Remote Homology Detection," *Nucleic Acids Res.* **30**(19), 4321–4328 (2002).

[8] E. Sonnhammer, S. Eddy, E. Birney, A. Bateman, and R. Durbin, "Pfam: MSAs and HMM-Profiles of Protein Domains," *Nucleic Acids Res.* **26**(1), 320–322 (1998).

[9] T. Doerks, R. Copley, J. Schultz, C. Ponting, and P. Bork, "Systematic Identification of Novel Protein Domain Families Associated with Nuclear Functions," *Genome Res.* **12**(1), 47–56 (2002).

[10] M. Madera, C. Vogel, S. Kummerfeld, C. Chothia, and J. Gough, "The SUPERFAMILY database in 2004: Additions and Improvements," *Nucleic Acids Res.* **32**(database issue), D235–D239 (2004).

[11] D. Wilson, M. Madera, C. Vogel, C. Chothia, and J. Gough, "The SUPERFAMILY Database in 2007: Families and Functions," *Nucleic Acids Res.* **35**(database issue), D308–D313 (2007).

[12] J. Gough and C. Chothia, "SUPERFAMILY: HMMs Representing All Proteins of Known Structure. SCOP Sequence Searches, Alignments and Genome Assignments," *Nucleic Acids Res.* **30**(1), 268–272 (2002).

[13] P. Papasaikas, P. Bagos, Z. Litou, V. Promponas, and S. Hamodrakas, "PRED-GPCR: GPCR Recognition and Family Classification Server," *Nucleic Acids Res.* **32**(web server issue), W380-W382 (2004).

[14] N. Sgourakis, P. Bagos, P. Papasaikas, and S. Hamodrakas, "A Method for the Prediction of GPCRs Coupling Specificity to G-Proteins Using Refined Profile Hidden Markov Models," *BMC Bioinformatics* **6**, 104 (2005).

[15] I. Alam, S. Hubbard, S. Oliver, and M. Rattray, "A Kingdom-Specific Protein Domain HMM Library for Improved Annotation of Fungal Genomes," *BMC Genomics* **8**, 97 (2007).

[16] M. Gollery, D. Rector, and J. Lindelien, "TLFAM–a New Set of Protein Family Databases," *OMICS* **6**(1), 35–7 (2002).

[17] C. Claudel-Renard, C. Chevalet, T. Faraut, and D. Kahn, "Enzyme-Specific Profiles for Genome Annotation: PRIAM," *Nucleic Acids Res.* **31**(22), 6633–6639 (2003).

[18] J. Pinney, M. Shirley, G. McConkey, and D. Westhead, "MetaSHARK: Software for Automated Metabolic Network Prediction from DNA Sequence and Its Application to the Genomes of *Plasmodium Falciparum* and *Eimeria tenella*," *Nucleic Acids Res.* **33**(4), 1399–1409 (2005).

[19] C. Hyland, J. Pinney, G. McConkey, and D. Westhead, "metaSHARK: a WWW Platform for Interactive Exploration of Metabolic Networks," *Nucleic Acids Res.* **34**(web server issue), W725–W728 (2006).

[20] J. Schultz, F. Milpetz, P. Bork, C.P. Ponting, "SMART, a Simple Modular Architecture Research Tool: Identification of Signaling Domains," *Proc. Natl. Acad. Sci. USA* **95**(11), 5857–5864 (1998).

[21] P. Thomas, A. Kejariwal, M. Campbell, H. Mi, K. Diemer, N. Guo, I. Ladunga, B. Ulitsky-Lazareva, A. Muruganujan, S. Rabkin, J. Vandergriff, and O. Doremieux, "PANTHER: a Browsable Database of Gene Products Organized by Biological Function, Using Curated Protein Family and Subfamily Classification," *Nucleic Acids Res.* **31**(1), 334–341 (2003).

[22] A. Marchler-Bauer, J. Anderson, M. Derbyshire, C. DeWeese-Scott, N. Gonzales, M. Gwadz, L. Hao, S. He, D. Hurwitz, J. Jackson, Z. Ke, D. Krylov, C. Lanczycki, C. Liebert, C. Liu, F. Lu, S. Lu, G. Marchler, M. Mullokandov, J. Song, N. Thanki, R. Yamashita, J. Yin, D. Zhang, and S. Bryant, "CDD: a Conserved Domain Database for Interactive Domain Family Analysis," *Nucleic Acids Res.* **35**(database issue), D237–D240 (2007).

[23] A. Marchler-Bauer, A. Panchenko, B. Shoemaker, P. Thiessen, L. Geer, and S. Bryant, "CDD: a Database of Conserved Domain Alignments with Links to Domain Three-Dimensional structure," *Nucleic Acids Res.* **30**(1), 281–283 (2002).

[24] A. Marchler-Bauer, J. Anderson, M. Derbyshire, C. DeWeese-Scott, N. Gonzales, M. Gwadz, L. Hao, S. He, D. Hurwitz, J. Jackson, Z. Ke, D. Krylov, C. Lanczycki, C. Liebert, C. Liu, F. Lu, S. Lu, G. Marchler, M. Mullokandov, J. Song, N. Thanki, R. Yamashita, J. Yin, D. Zhang, and S. Bryant, "CDD: a Conserved Domain Database for Protein Classification," *Nucleic Acids Res.* **33**(database issue), D192–D196 (2005).

[25] A. Marchler-Bauer, J. Anderson, M. Derbyshire, C. DeWeese-Scott, N. Gonzales, M. Gwadz, L. Hao, S. He, D. Hurwitz, J. Jackson, Z. Ke, D. Krylov, C. Lanczycki, C. Liebert, C. Liu, F. Lu, S. Lu, G. Marchler, M. Mullokandov, J. Song, N. Thanki, R. Yamashita, J. Yin, D. Zhang, and S. Bryant, "CDD: a Curated Entrez Database of Conserved Domain Alignments," *Nucleic Acids Res.* **31**(1), 383–387 (2003).

[26] R. Tatusov, N. Fedorova, J. Jackson, A. Jacobs, B. Kiryutin, E. Koonin, D. Krylov, R. Mazumder, S. Mekhedov, A. Nikolskaya, B. Rao,

S. Smirnov, A. Sverdlov, S. Vasudevan, Y. Wolf, J. Yin, and D. Natale, "The COG Database: an Updated Version Includes Eukaryotes," *BMC Bioinformatics* **4**, 41 (2003).

[27] R. Tatusov, D. Natale, I. Garkavtsev, T. Tatusova, U. Shankavaram, B. Rao, B. Kiryutin, M. Galperin, N. Fedorova, and E. Koonin, "The COG Database: New Developments in Phylogenetic Classification of Proteins from Complete Genomes," *Nucleic Acids Res.* **29**(1), 22–28 (2001).

[28] R. Tatusov, M. Galperin, D. Natale, and E. Koonink, "The COG Database: a Tool for Genome-Scale Analysis of Protein Functions and Evolution," *Nucleic Acids Res.* **28**(1), 33–36 (2000).

Chapter 5

Building an Analytical Pipeline

5.1 INTRODUCTION

Building an analytical pipeline to manage your workflow processes can save time and improve consistency by establishing a set automated routine that is followed every time. Bioinformaticists frequently utilize HMM databases as part of their pipelines. Chapter 5 refers to the development and use of HMM databases in an automated bioinformatics workflow. There are at least three ways to set up an analytical pipeline on your own server:

1. Write your own from scratch.

2. Use a workflow tool, whether from a commercial group or open source.

3. Use an existing pipeline, such as InterProScan or PartiGene.

This chapter will discuss all three methods, including ways to modify InterProScan to include custom databases.

5.2 WHAT IS AN ANALYTICAL PIPELINE?

Most biologists and bioinformaticists tend to do many of the same processes repeatedly across many datasets. One common pipeline process might be as follows: importing chromatograms from the sequencer, reading the bases, masking the vector data, removing the repeat regions, clustering and assembling the data into contigs, and finally running hmmPfam on the contigs. If this process is to be used for a number of projects, a pipeline should be created that will automatically tie the required programs together. In this way, the output of one program will automatically feed into the input of the next program, as shown in figure 5.1.

A pipeline saves both researcher time and overall CPU time. The savings in researcher time is obviously because the data flow from one function to

From This: to This:

Copy Chromats

Read Bases

Mask Vector Data

Remove repeats

Cluster

Assemble

Search against PFAM

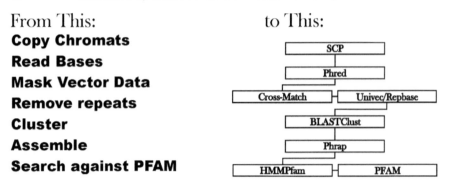

FIGURE 5.1:   An Example Pipeline

the next with no input required other than the initial parameter establishment. The savings in computer time comes from more efficient utilization of resources—if one step in the process finishes in the middle of the night, the next one starts without having to wait for someone to arrive and issue the next command.

Depending on the algorithms in use, the pipeline may be able to further improve performance by overlapping processes in a parallel manner. That is, as one algorithm generates output, the next begins work on the files as they are generated, and proceeds processing. Some studies have shown a speedup of nine times over a serial processing scheme, but it is important to note that this is not always an appropriate scheme for every situation. For example, clustering must be entirely completed before the next step can be started.

Some people differentiate a workflow system from a pipeline in terms of complexity. A pipeline simply describes the flow of data through a series of processes, while a workflow process can branch the flow of data based on a series of rules. For example, if the datasets are BLASTed against the Uniref90 database, then the results might be parsed into two bins, one with sequences that had hits to uniref90, and one with sequences that have no hits at that threshold. These two groups of sequences may be treated differently; the group with no hits could be run against Pfam, for example. Then those that still found no hits with Pfam could be run against one or more of the other HMM databases mentioned in chapter 4, such as PANTHER.

Others use the term pipeline interchangeably with workflow management system, and in practice we often use the same tools whether the system is simple and straightforward or complex and branching. We will use both terms in this discussion, particularly since pipelines tend to become more complex over time.

Many systems glue together programs and data that reside solely on a local workstation or server, but others will use remote services to some degree. In this scenario, the script submits the data to a web server that provides some service—the NCBI BLAST server is popular for this—and then waits for the results to be returned before going on to the next step. The advantages of this type of pipeline are that the user avoids the setup and maintenance issues of running the program locally, and the compute cycles used are effectively donated by someone else. The disadvantages are that the amount of data that may be processed is frequently much lower than on your own computer, and it can be difficult to predict how long something will take. If someone else submits a huge job just before you submit yours, then your results may be delayed, sometimes for days.

5.3 HOW DO I CREATE A PIPELINE, AND WHAT DO I PUT INTO IT?

Bioinformatics relies heavily on the analysis of sequences, and as a result a lot of text string manipulation is required. Perl is the standard language for generating pipelines, partly because it is an excellent language for handling text strings. Many people prefer other languages for one reason or another (Python and Ruby are also popular), but overall Perl has the greatest impact in the bioinformatics community.

Let's say a researcher has a group of files, each one containing multiple related sequences. The files are named with a number, seq1.faa, seq2.faa.....seq*n*.faa. If there are only a few files, then they may be aligned with ClustalW or MUSCLE, built into HMMs and calibrated by issuing the following three commands for each of the files:

>clustalw –infile=seq1.faa –ktuple=5 –window=10 –outputtree

=phylip –outfile=seq1.aln

>hmmbuild –f seq1.hmm seq1.aln

>hmmcalibrate –cpu 2 seq1.hmm

If the number of these files is more than just a few, the researcher may find this to be a rather tiresome chore. Additionally, if each file holds a good number of sequences to align, then the time required may become rather annoying, and we would greatly prefer to let it run all by itself while we go have coffee.

For a simple task like this, a simple shell script would suffice, but these things have a way of getting complicated quickly, so we might as well start in Perl. For example, we might find that some of our files have duplicate sequences, and clustalw will give us a message that they need to be cleaned up before the alignments can be created. Fortunately, Paul Stothard of the Canadian bioinformatics help desk has already written a Perl script to remove duplicates, so we simply need to add this to our process and we may continue on with our work.

The procedure now flows like this:

remove_duplicates → build alignments → build HMMs → calibrate HMMs

If we have a few files to process, we may simply paste the commands in the script repeatedly, change seq1 to seq$n$ where n is the number of the file that we are using, and that will work well. In practice, the number of sequences will quickly exceed our patience, and we will need to include a 'foreach statement.' This will make our script much more maintainable and flexible, not to mention shorter.

Many arguments have been made about the type of scripting language to use. These debates generally turn into something that resembles a religious war. I do not have a particular bias toward Perl in and of itself; personally, I like Python, and I hear wonderful things about Ruby as a language. For the purpose of gluing together applications into a pipeline, Perl has the advantage simply because of the number of tools that are already out there, freely available. The fastest way to write a script is to download it off the Internet!

Perl also has an advantage in that it is available for virtually any operating system. Setup is easy, no matter which type of computer you use. Converting scripts (for example, from Linux to Windows) can generally be done by changing the directory references.

Several good books are available to teach Perl for bioinformatics. James Tisdall has written the most popular, entitled *Beginning Perl For Bioinformatics*, and the follow-up book, *Mastering Perl For Bioinformatics*. These will provide the necessary background for producing a set of workable scripts, or editing an existing pipeline.

Some of the pipeline systems mentioned in section 5.5 are themselves created from a series of scripts. Therefore, it is a relatively simple matter to add, subtract or replace components to create an improved pipeline to meet the needs of your own specific project. For example, if a system calls for a search against the Pfam database, it is an easy matter to replicate that section of code to repeat the search with some of the other databases mentioned in chapter 4, such as TIGRfam, SuperFamily, and SMART as in figure 5.2.

Depending on the objectives of the project, the branching aspect of a workflow can save a considerable amount of time in the construction of a 'data distillation column' as in figure 5.3. The sequences could first be BLASTed against a well-annotated database such as SwissProt, and then only the sequences that have no hits could be run against Pfam. Since BLAST is a much faster algorithm, this will reduce the runtime considerably. Then the sequences that still have no hits can be run against other HMM databases, like SuperFamily, TigrFam, and SMART, for example.

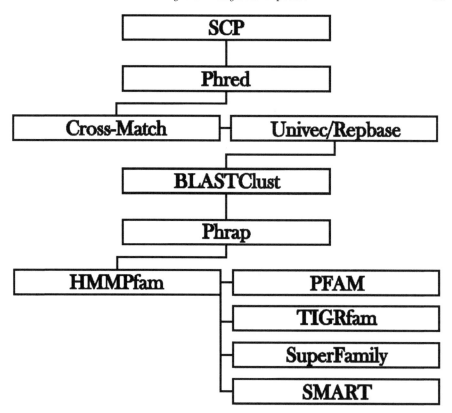

FIGURE 5.2: Pipeline with Additional HMMPfam Searches Added

While Perl scripts may be configured to work on any computer system, the executable programs for bioinformatics are typically Linux-based. There are certainly some notable exceptions, but, in general, most bioinformatics programs run under Linux. Even if you are running under some version of Linux, however, you may not wish to run all of the programs of your pipeline on your own workstation. Therefore, it may be advantageous to design your workflow to issue commands to a remote server to enable execution to take place at the best possible location, whether this is to obtain access to a higher performance system, cluster, or grid, or whether it is to run the programs on the appropriate operating system.

Running the workflow on a remote server may also be useful, simply because the necessary programs are available on that server, and not on the local machine.

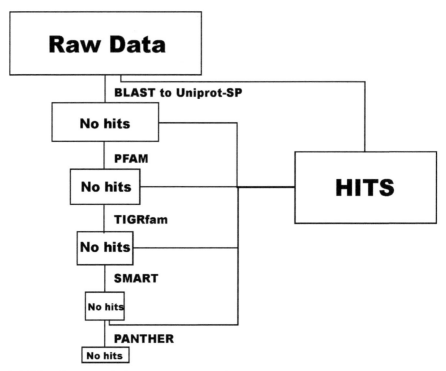

FIGURE 5.3: A Data Distillation Column

5.4 IS THERE AN EASIER WAY TO MANAGE MY WORKFLOW?

Bioinformatics analysis pipelines have a way of growing far beyond the original design goals and parameters. Either the goals are broadened dramatically or the volume of data is far beyond what was originally envisioned. Sometimes both situations occur for the same project!

Building a simple script for managing repetitive work is a good thing, and can really simplify your life. Maintaining a set of Perl scripts in a rapidly changing or expanding project, however, can become impossible. Programmers that start this type of project have no problem with this, because they are now guaranteed a good salary for life, but for those of us who need to publish papers and finish projects, a lack of scalability is a very bad thing!

Fortunately, a number of programs have arisen that can quickly generate even the most complex workflow with little to no programming effort. The modification, expansion and testing of the resulting programs is accomplished with little of the anguish that can arise with a set of scripts, that always seem to need just one more little bug to be fixed!

Some of the development tools listed here are commercial, such as VIBE and Pipeline Pilot. But if these tools can save the cost of a programmer's salary, then the cost to buy the software may well be worth the investment.

Below are some of the commercial tools.

Pipeline Pilot/Accelrys Pipeline Pilot has a drag-and-drop interface for building applications with a minimum of programming effort. Many other companies, such as Tripos and Molecular Networks, have developed components for the pipeline pilot system, so that their software may easily work in the pipelines that are generated. Pipeline Pilot has partnered with Altair Engineering so that the data analyses will be easily able to spread over clusters and grids running PBS.

VIBE/Incogen The Visual Integrated bioinformatics Environment (VIBE) interfaces with a number of high throughput platforms, such as the Sun Grid Engine (SGE) and the TimeLogic DeCypher, for the highest performance. VIBE has more flexibility than some of the other systems here in that the architecture may be client/server or simply client side (that is, running on the desktop). VIBE uses XML for communication between processes, data storage and data exchange. A Java API is provided so that users can write their own modules.

Biosense/Inforsense Biosense is built on the Inforsense KDE (Knowledge Discovery Environment). Biosense is designed to allow end users to rapidly create and customize their own analytical applications in an interactive fashion. Biosense fully integrates Perl and shell scripts and programs written in R/Bioconductor and Matlab. The Biosense toolbox includes a dendrogram viewer, SNP analytical tool, expression analysis tool, sequence analysis tool, and proteomics system. Biosense is built to work with FASTA, BLAST, EMBOSS, ClustalW and more.

PipeWorks PipeWorks is primarily designed to be used on a DeCypher system from TimeLogic. Other tools may be added at any time, whether from TimeLogic or from an outside source.

Open-Source Pipeline Development Tools One of the advantages to buying a commercial system is that there is a built-in support system that can at least theoretically fix some of the bugs that arise. On the other hand, an open-source system can be more easily distributed to your collaborators without fear of violating the licensing agreement.

Taverna Taverna is the most widely used of the workflow environment tools. Taverna was built to work with the MyGrid project, and is a product of the EBI and several universities. Taverna will run on Windows, Mac, Linux and most UNIX-like operating systems. The interoperability of Taverna is excellent, as it works well with BioMoby, Pfam, EMBOSS, and BioConductor, as well as many others.

Taverna uses web services as components. SOAPLAB is used to provide web services from command-line applications. The user connects the input and output ports of the components to create a workflow.

Dozens of workflows are already available to work with Taverna. Tutorials are available online as well as documentation to help you get started.

Biopipe Biopipe is an excellent tool that is designed to execute pipelines across clusters. It is not capable of generating more complex workflows with branches and iterations, however. The interface is a bit more complex than some of the other systems that provide a graphical means for generating designs.

Wildfire In a move away from the web services based applications such as Taverna, Wildfire has direct access to the executables. Rather than generating Perl scripts, Wildfire executes the workflows using GEL (Grid Execution Language). GEL is capable of running the executables directly, over the compute nodes of a cluster, or on the grid using Condor.

Frequently, a set of Hidden Markov Models is tested for sensitivity by leaving out some fraction of the training data (say 10%), and then using that portion to test the efficacy of the models. Then the process is repeated with a separate 10% being held out, and so forth. The authors of Wildfire have demonstrated how their tool can automate this process and therefore reduce a tedious chore considerably, using an allergen database as an example.

GenePattern The Broad Institute and MIT have produced the GenePattern tool with different goals in mind compared to some of the others listed here. Rather than emphasize a front end to EMBOSS, GenePattern is designed with gene expression analysis and proteomics in mind, and provides integration for BioConductor and geWorkbench.

5.4.1 What to Look for in a Workflow Development Tool

Be sure that whichever system that you choose is able to handle errors well—what happens if one of the programs does not complete, either because of a network outage or a program crash? You should be able to access the error logs easily, to see what has happened.

The workflow should be portable to other systems. The open-source tools have the edge here. If you build a workflow system that works well for you, you ought to be able to share it easily with others. Best of all are the network services type of pipelines, because they can still point to the same locations no matter where they may be running. Making the system portable is crucial, because even though these pipeline tools are much easier than programing, it is still common for one person to create the workflow for someone else to run on a regular basis.

It should be easy to include new components into the system. bioinformatics programs are continuously being developed, and you should be able to incorporate them into your workflow with a minimum of bother.

5.5 ARE THERE ANY PIPELINES THAT I CAN SIMPLY DOWNLOAD AND INSTALL?

While the workflow development tools mentioned in section 5.4 are quite useful, and much easier even than Perl, I always prefer to let someone else do the work whenever possible! Some of the many pipelines that are written in Perl include

- InterProScan, which has become extremely popular for new genome annotation.

- PartiGene, which is designed for EST analysis and transcript reconstruction.

- PDA, Pipeline Diversity Analysis, which automatically searches for polymorphic sequences

- BOD, bioinformatics On Demand, pays particular attention to parallelism on the grid.

- ESTannotator has a nicer viewer for BLAST hits than other systems, but it does not include submission to DbEST.

- MAGIC-SPP includes a Java front end, an Oracle database, and a LIMS system, for an overall package that is more robust and scalable than most.

While each of these packages is valuable in many ways, we will take a more in-depth look at the first two—InterProScan and PartiGene.

5.5.1 InterProScan

InterPro (http://www.ebi.ac.uk/interpro) is a combination of a number of signature databases. Curated at the EBI, it is built from Pfam, Prints, Prosite, ProDom, SMART, SuperFamily, TIGRFAMs, PANTHER, PIRsf, Gene3D and SP/TrEMBL. This is an astonishingly powerful tool that is free to download. InterPro entries include families, domains, repeats, active sites, binding sites and post-translational modification sites.

While InterPro is a useful database, the website only allows the analysis of a single sequence at a time. To analyze large amounts of data (say, an entire Proteome) one needs to install InterProScan. InterProScan is the set of programs and scripts needed to compare your sequences against InterPro on your own server. InterProScan is essentially a free analysis pipeline that is flexible, extensible and very powerful. Large analysis jobs are split up and spread over a cluster, and then reassembled at the end. SGE, PBS and LSF are supported. Output can be generated in text, Tab, XML or HTML format. A web interface is provided. GO mappings can be automatically assigned.

The heavy use of HMM databases tends to make searches very slow. Analyzing a genome of 30,000 genes can take two months of CPU time! Changing out the HMMPfam executable for ldHMMer or one of the other optimized versions might drop this down to a few days, and would most likely be much cheaper than expanding the server farm. Moving the HMMPfam search to an accelerator could reduce the time down to a few hours, which would more than compensate for the time to write the appropriate scripts.

While InterPro already incorporates an impressive array of HMM databases, you may wish to add your own models or even entire collections of models to the mix. There are at least two ways to do this. Since InterProScan is written in Perl, it is a relatively simple matter to edit the scripts to include your database in the search. Unless you can provide links from your database to IPR numbers, however, you may not get the complete output that you expect.

A simpler method to include your data into the InterProScan is to simply concatenate your data into one of the existing databases, such as Pfam. This is especially useful if you only have a few models to add.

An alternate scenario is also possible. If you are only searching with the intent to find a certain subset of the data, not the full selection, then why wait while InterProScan searches the entire database? Either edit the script so that not all databases are searched, or edit the databases themselves to remove the models that are not of interest. The resulting search will be faster depending upon the amount of reduction.

5.5.2   InterPro Content

While the content changes rapidly, the information on the makeup of the InterPro database is provided in figure 5.4 to give an impression of the relative contribution of the various member data sources. InterPro release 15.0 contains nearly 15,000 entries, representing 4454 domains, 10006 families, 231 repeats, 34 active sites, 20 binding sites and 19 post-translational modification sites. Overall, there are over 17,000,000 InterPro hits from the UniProtKB protein sequences. A complete list is available from the InterPro ftp site. The details may be viewed graphically as shown in figure 5.5.

DATABASE	VERSION	ENTRIES
PANTHER	6.1	30128
Pfam	21.0	8957
PIRSF	2.68	1748
PRINTS	38.0	1900
ProDom	2005.1	1522
PROSITE	20.0	2006
SMART	5.0	706
TIGRFAMs	6.0	2946
GENE3D	3.0.0	2147
SUPERFAMILY	1.69	1538
UniProtKB/Swiss-Prot	52.0	261513
UniProtKB/TrEMBL	35.0	3987044
InterPro	15.0	14764
GO Classification	N/A	23937

FIGURE 5.4: InterPro Entries

InterPro domain architecture:

InterPro Entry	Method accession	Graphical match ☑	Method name
IPR000242:	PF00102		V_phosphatase
IPR000242:	PR00700		PRTVPHPHTASE
IPR000242:	PS50055		TYR_PHOSPHATASE_PTP
IPR000242:	SM00194		PTPc
IPR000387:	PS00383		TYR_PHOSPHATASE_1
IPR000387:	PS50056		TYR_PHOSPHATASE_2
IPR000980:	PD000093		SH2
IPR000980:	PF00017		SH2
IPR000980:	PR00401		SH2DOMAIN
IPR000980:	PS50001		SH2
IPR000980:	SM00252		SH2

Classification	PDB Chain/Domain ID & View 3D	PDB Chain/Structural Domains ☑	
2shp	2shpa		
2shp	2shpb		
3.30.505.10.1	2shpA1		
3.30.505.10.1	2shpA2		
3.90.190.10.2	2shpA3		
c.45.1.2	c2shpa1		
d.93.1.1	c2shpa2		
d.93.1.1	c2shpa3		

Figure 1. Illustration of the detailed view for protein Q06124, the human protein-tyrosine phosphatase, non-receptor type 11. From an InterPro entry page, clicking on a protein accession number in the 'Examples' field takes you to this view for that protein. The oval shapes at the top of the figure display the InterPro Domain Architecture (IDA) view for this protein, which represents its domain composition. Each oval shape contains the domain name and the number of its iterations of the domain if greater than one. The InterPro detailed view represents the protein sequence as a series of different lines for each protein signature hit. The bars are colour coded according to the member database. A separate view below the signature matches displays the structural domains from the SCOP and CATH as white-striped bars. This view provides a complete picture of the protein domain composition and where sequence-based domains correspond to known structures.

FIGURE 5.5: InterPro Detailed View. Domain architecture is at the top. The graphical match view represents the protein as a number of lines for each signature hit. The different databases are represented by different colored bars. Structural domains from SCOP and CATH are shown below as white-striped bars.

5.5.3   PartiGene

The Blaxter lab at the University of Edinburgh has developed a series of tools to process EST data for the genomes of nematodes, but many others have found that these are generally applicable to a wide variety of organisms. PartiGene is easy to install, extend, modify and manage. The following components are available as part of the total system:

Trace2DbEST takes sequence data from the chromatogram stage, calls the bases, cleanses the sequence, provides a BLAST-based annotation and submits the appropriate data to NCBI. A module called CLOBB does the clustering.

PartiGene clusters and assembles the sequences, then BLASTs the clusters and enters the data into a PostgreSQL database.

The wwwPartiGene program provides a web interface to the Partigene database.

Prot4EST translates the cleansed, clustered and assembled data into protein sequence through a pipeline that provides a robust system for translation. Six steps are followed to ensure the highest quality. Ribosomal RNA is identified and not translated. Mitochondrial ESTs are identified and then translated using the mitochondrial genetic code. ORFs are compared to a protein database to build a minimal tiling path translation of the EST. Next, ESTscan is used to identify open reading frames. Then DECODER is used to predict peptides. Finally, if none of these yield a peptide prediction, one is made from the longest ORF in the six reading frame predictions.

Annot8r adds Gene Ontology (GO) annotations, EC numbers and KEGG pathways based on BLAST similarity searches, and these annotations are stored in the database.

Annot8r_physprop calculates a number of physical properties for the predicted peptide sequences, along with information on the reliability of these calculations.

Another system called Paeano builds upon the PartiGene system and adds InterProScan, OrthoLog (which finds orthologs in model organisms) and GEOExpress (which adds Gene Expression data).

5.6   SUMMARY

In the mid-1980's, personal computing made the transition from a small hobbyist fad to a widespread phenomenon. Eventually, enough systems were sold that the personal computer became ubiquitous in business, effectively eliminating the mainframes that had previously dominated the business world.

The real power of these systems, however, did not become obvious until enough of them were connected together via networking. Similarly, the real power of bioinformatics algorithms will only truly be realized when they are easily connected to each other so that the programs themselves become like nodes in a network. Repetitive tasks will be made much more bearable, and analyses will become more reproducible through the use of workflow systems.

Workflows may be generated in one of three ways:

1. Through the use of hand-coded scripts, typically written in Perl.

2. As a product of an automatic workflow generating tool.

3. As a downloaded, prebuilt project that has been generated by a commercial or open-source software project. This type of system may be modified as needed to fit the goals and objectives of the project through the modification of the underlying scripts.

Each of these methods has its own advantages and disadvantages, and the method chosen will differ depending upon the goals of the project, the amount of time available and the skills of the participants.

5.7   QUESTIONS

1. A project has just generated 50,000 ESTs that must be processed and entered into DbEST quickly prior to a grant proposal. Discuss how you would go about this and why.

2. The complete set of protein sequences from ten model organisms must be analyzed in terms of Gene Ontology categories. How would you go about doing this? What challenges would there be in the implementation of your approach?

3. If there are several already existing pipelines available to meet a wide range of needs, why would it still be necessary for a bioinformaticist to learn Perl?

4. Download any workflow design tool listed here and install it. Build a simple workflow that BLASTs a set of protein sequence data against Uniref90, runs an HMMPfam search against the Pfam database and then compares the gi numbers of the sequences that found no hits to Pfam to those that found no hits to Uniref.

5. Running InterProScan locally can have a great speed advantage over the Interpro website when analyzing a large amount of data. Visit the Interpro website and give three advantages of using the website over running the searches locally.

6. Most analytical pipelines utilize an open-source database such as MySQL to store the data. Some, however, use Oracle. What advantages and disadvantages might there be to each of these approaches?

7. Compiled languages tend to be much faster than interpreted languages. What reasons are there to not use a compiled language like C++ to construct a pipeline?

8. Install any two workflow development packages on your system. Compare and contrast the installation procedures.

9. Install any two open-source pipelines on your system. Compare and contrast the installation procedures.

References

[1] J. Wasmuth and M. Blaxter, "prot4EST: Translating Expressed Sequence Tags from Neglected Genomes," *BMC Bioinformatics* **5**, 187 (2004).

[2] C. Liang, F. Sun, H. Wang, J. Qu, R. Freeman, L. Pratt, and M-M Cordonnier-Pratt, "MAGIC-SPP: a Database-Driven DNA Sequence Processing Package with Associated Management Tools," *BMC Bioinformatics* **7**, 115 (2006).

[3] P. Shafer, D. Lin, and G. Yona, "EST2Prot: Mapping EST Sequences to Proteins," *BMC Genomics* **7**, 41 (2006).

[4] Y. Strahm, D. Powell, and C. Lefvre, "EST-PAC a Web Package for EST Annotation and Protein Sequence Prediction," *Source Code Biol Med.* **1**, 2 (2006).

[5] J. Peregrn-Alvarez, A. Yam, G. Sivakumar, and J. Parkinson, "PartiGeneDB—Collating Partial Genomes," *Nucleic Acids Res.* **1**, 33(database issue), D303–D307 (2005).

[6] D. Field, B. Tiwari, and J. Snape, "Bioinformatics and Data Management Support for Environmental Genomics," *PLoS Biol.* **3**(8), e297 (2005).

[7] M. Latorre, H. Silva, J. Saba, C. Guziolowski, P. Vizoso, V. Martinez, J. Maldonado, A. Morales, R. Caroca, V. Cambiazo, R. Campos-Vargas, M. Gonzalez, A. Orellana, J. Retamales, and L. Meisel, "JUICE: a Data Management System that Facilitates the Analysis of Large Volumes of

Information in an EST Project Workflow," *BMC Bioinformatics* **7**, 513 (2006).

[8] M-M Cordonnier-Pratt, C. Liang, H. Wang, D. Kolychev, F. Sun, R. Freeman, R. Sullivan, and L. Pratt, "MAGIC Database and Interfaces: an Integrated Package for Gene Discovery and Expression," *Comp. Funct. Genom.* **5**, 268–275 (2004).

[9] F. Tang, C. Chua, L. Ho, Y. Lim, P. Issac, and A. Krishnan A. "Wildfire: Distributed, Grid-Enabled Workflow Construction and Execution," *BMC Bioinformatics* **6**, 69 (2005).

[10] P. Romano, D. Marra, and L. Milanesi, "Web Services and Workflow Management for Biological Resources," *BMC Bioinformatics* **6**(Suppl 4), S24 (2005).

[11] E. Quevillon, V. Silventoinen, S. Pillai, N. Harte, N. Mulder, R. Apweiler, and R. Lopez, "InterProScan: Protein Domains Identifier," *Nucleic Acids Res.* **33**(web server issue): W116–W120 (2005).

[12] N. Mulder et. al, "New developments in the InterPro Database," *Nucleic Acids Res.* **35**(Database issue):D224–8 (2007).

[13] E. Zdobnov and R. Apweiler, "InterProScan–an Integration Platform for the Signature-Recognition methods in InterPro," *Bioinformatics* **17**(9), 847–8 (2001).

[14] S. Casillas and A. Barbadilla, "PDA: a Pipeline to Explore and Estimate Polymorphism in Large DNA Databases," *Nucleic Acids Res.* **32**(web server issue), W166–W169 (2004).

[15] S. Casillas and A. Barbadilla, "PDA v.2: Improving the Exploration and Estimation of Nucleotide Polymorphism in Large Datasets of Heterogeneous DNA," *Nucleic Acids Res.* **34**(web server issue), W632–W634 (2006).

[16] L. Qiao, J. Zhu, Q. Liu, T. Zhu, C. Song, W. Lin, G. Wei, L. Mu, J. Tao, N. Zhao, G. Yang, and X. Liu, "BOD: a Customizable Bioinformatics on Demand System Accommodating Multiple Steps and Parallel Tasks," *Nucleic Acids Res.* **32**(14), 4175–4181 (2004).

[17] E. Kawas, M. Senger, and M. Wilkinson, "BioMoby Extensions to the Taverna Workflow Management and Enactment Software," *BMC Bioinformatics* **7**, 523 (2006).

[18] M. Hassan, R. Brown, S. Varma-O'Brien, and D. Rogers, "Cheminformatics Analysis and Learning in a Data Pipelining Environment," *Mol. Divers* **10**(3), 283–99 (2006).

[19] R. de Knikker, Y. Guo, J. Li, A. Kwan, K. Yip, D. Cheung, and K. Cheung, "A Web Services Choreography Scenario for Interoperating Bioinformatics Applications," *BMC Bioinformatics* **5**, 25 (2004).

[20] A. Konagaya, "Trends in Life Science Grid: From Computing Grid to Knowledge Grid," *BMC Bioinformatics* **7**(Suppl 5), S10 (2006).

[21] R. de Knikker, Y. Guo, J. Li, A. Kwan, K. Yip, D. Cheung, and K. Cheung, "A Web Services Choreography Scenario for Interoperating Bioinformatics Applications," *BMC Bioinformatics* **5**, 25 (2004).

[22] A. Garcia Castro, S. Thoraval, L. Garcia, and M. Ragan, "Workflows in Bioinformatics: Meta-Analysis and Prototype Implementation of a Workflow Generator," *BMC Bioinformatics* **6**, 87 (2005).

[23] D. Hull, K. Wolstencroft, R. Stevens, C. Goble, M. Pocock, P. Li, and T. Oinn, "Taverna: a Tool for Building and Running Workflows of Services," *Nucleic Acids Res.* **34**(web server issue), W729–W732 (2006).

[24] E. Kawas, M. Senger, and M. Wilkinson, "BioMoby Extensions to the Taverna Workflow Management and Enactment Software," *BMC Bioinformatics* **7**, 523 (2006).

Chapter 6

Building Custom Databases

6.1 INTRODUCTION

Chapter 6 is a practical guide for making custom HMM databases using the HMMer, SAM and PSI-BLAST packages. Tips and tricks from personal experience will be included, as will links to some online resources to make the process less painful.

Before building your own database, it is important to have a clear objective in mind as to the intended outcome of the project. Why do you want to build a custom HMM collection when there are so many excellent resources available already? There are several possible reasons why this might be desirable. The first is that you may have a group or several groups of related proteins that are not yet represented in the existing databases. You may have expert knowledge about some types of proteins and can cluster them in a way that has not been done before.

On the other hand, you may find that you want to produce models with a greater specificity than the public databases, which tend to be more general. This is similar to what has been done with PRED-GPCR. It is easy to identify that a sequence is a GPCR through a BLAST search, but the type of GPCR is not so clear. PRED-GPCR can identify this easily because it is focused on this one particular aspect of GPCR classification. On the other hand, the FPfam database does not have finer-grain detail about the range of proteins, but instead has more specific models for the type of organisms—in this case, fungal organisms.

Once the database is built, a side benefit may be that since the database is usually smaller as a result of being more focused, searches will be much faster.

There are several approaches that one may take to create a special-purpose HMM database. The process that follows is a simplified example that is given for illustrative purposes, and may not suit your particular needs. See the documentation of the databases already mentioned to get additional tips and techniques.

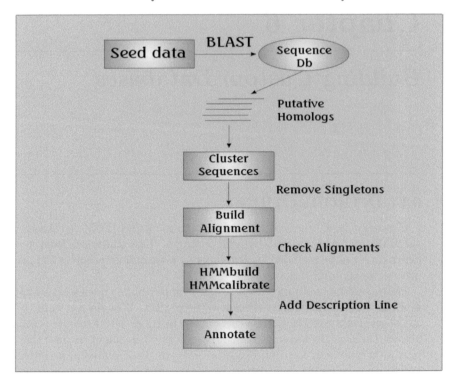

FIGURE 6.1:   An Overview of the Database Building Process

6.2   BUILDING HMMER DATABASES

One example of a process to generate Hidden Markov Models is shown in figure 6.1. Other methods may involve variable beginning points, for example, if the seed alignments are already made, then the initial steps may be bypassed.

The method described here begins with a BLAST search of a sequence database. The resulting alignments are extracted, aligned with each other using ClustalW or some other tool, and built into a Hidden Markov Model. Another method might start with a model or group of models from Pfam, and generating the initial alignments with hmmsearch. Others may start with a group of validated sequences that represent a known family of proteins.

First, BLAST the sequences that you will use for your 'seeds' as shown in figure 6.2.

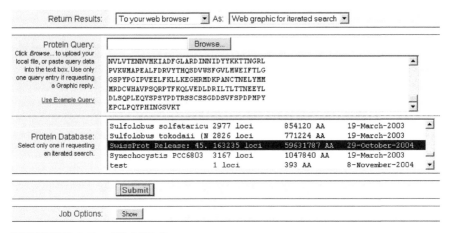

FIGURE 6.2: BLAST Step

Select the hits that you will use for your model as shown in graphical form in figure 6.3. Notice that simply grabbing the top hits would not make this a very general model, because they are too similar to one another. To maximize entropy in the model, select a good representation of all the hits. This will produce a model that is better able to detect remote homologues to the protein family.

Build the Multiple Sequence Alignment (MSA) as shown in figure 6.4.

Using HMMbuild, create the model, and then calibrate it with HMMCalibrate

Using that model, search the database using hmmsearch. This should find more hits that may then be incorporated into the model. This step may be iterated until no new hits are found.

6.2.1 Calibration of the Model

When using the HMMer package, the scoring of the model is not really optimal until the model is calibrated. The HMMcalibrate function does this by running a large amount of random data through the model. A histogram is generated out of those scores, and an extreme value distribution is fitted to that histogram. The model is saved with those EVD parameters.

The HMMcalibrate function is slow, typically several minutes per model depending on length and the speed of the CPU used. There is no optimized or accelerated version of HMMcalibrate, at the time of this writing. It is best to concatenate the models into one big flatfile database, and then calibrate them all overnight, or over the weekend, depending on the size of the database and the speed of your computer.

If a set of models are used that have not been calibrated, the job will still run, and scores will be obtained. The resultant E-values, however, will not

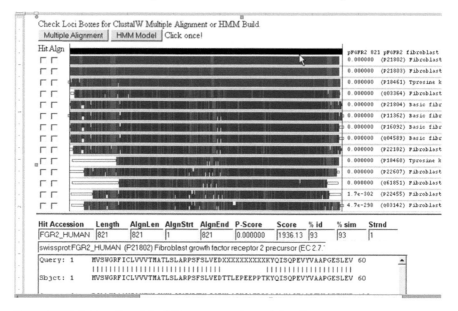

FIGURE 6.3: Graphical Representation of the Output

be trustworthy, and if they are used in close comparisons, the level of errors may be excessively high. Studies have found that the statistics produced by HMMer are quite reliable (figure 6.5 shows an example of a portion of a model), but this is only if this step is in place. If the models are not properly calibrated, then those studies do not apply.

6.3 BUILDING DATABASES WITH THE SAM PACKAGE

While the SAM and HMMer packages are very similar conceptually, a good deal of difference is experienced with the actual implementation. Perhaps the most obvious difference to the user is that SAM models do not require calibration. Building SAM databases is made considerably easier through the use of the web-based SAM-T06 program.

SAM-T06 is an extremely interesting tool; it is extremely powerful and has no equivalent in the HMMer package. Earlier versions were dated by the year, such as T98, which was released in 1998, T99 which was released in 1999 and so forth.

The benefit of SAM-T06 is that one may start with a single sequence and wind up with a model, MSA, and structural predictions as well as the top hits to the model. On the downside, the website will only accept relatively small amino acid sequences in order to save computational resources. For larger

```
pFGFR2                       NGKEFKQEHRIGGYKVRNQHWSLIMESVVPSDKGNYTCVVENE
swissprot|FGR2_HUMAN|        NGKEFKQEHRIGGYKVRNQHWSLIMESVVPSDKGNYTCVVENE
swissprot|CEK3_CHICK|        NGKEFKQEHRIGGYKVRNQHWSLIMESVVPSDKGNYTCIVENC
swissprot|FGR1_CHICK|        NGKEFKPDHRIGGYKVRYATWSIIMDSVVPSDKGNYTCIVENK
swissprot|FGR1_RAT|          NGKEFKPDHRIGGYKVRYATWSIIMDSVVPSDKGNYTCIVENE
swissprot|CEK2_CHICK|        NGKEFKGEHRIGGIKLRHQQWSLVMESVVPSDRGNYTCVVENK
swissprot|FGR4_HUMAN|        DGQAFHGENRIGGIRLRHQHWSLVMESVVPSDRGTYTCLVENA
swissprot|FGR4_MOUSE|        DGQAFHGENRIGGIRLRHQHWSLVMESVVPSDRGTYTCLVENS

pFGFR2                       YHLDVVERSPHRPILQAGLPANASTVVGGDVEFVCKVYSDAQF
swissprot|FGR2_HUMAN|        YHLDVVERSPHRPILQAGLPANASTVVGGDVEFVCKVYSDAQF
swissprot|CEK3_CHICK|        YHLDVVERSPHRPILQAGLPANASAVVGGDVEFVCKVYSDAQF
swissprot|FGR1_CHICK|        YQLDVVERSPHRPILQAGLPANKTVALGSNVEFVCKVYSDPQF
swissprot|FGR1_RAT|          YQLDVVERSPHRPILQAGLPANKTVALGSNVEFMCKVYSDPQF
swissprot|CEK2_CHICK|        YQLDVLERSPHRPILQAGLPANQTVVVGSNVEFHCKVYSDAQF
swissprot|FGR4_HUMAN|        YLLDVLERSPHRPILQAGLPANTTAVVGSDVELLCKVYSDAQF
swissprot|FGR4_MOUSE|        YLLDVLERSPHRPILQAGLPANTTAVVGSDVELLCKVYSDAQF

pFGFR2                       VEKNGSKYGPDGLPYLKVLKAAGVNTTDKEIEVLYIRNVTFED
swissprot|FGR2_HUMAN|        VEKNGSKYGPDGLPYLKVLKAAGVNTTDKEIEVLYIRNVTFED
swissprot|CEK3_CHICK|        VERNGSKYGPDGLPYLQVLKAAGVNTTDKEIEVLYIRNVTFED
swissprot|FGR1_CHICK|        IEVNGSKIGPDNLPYVQILKTAGVNTTDKEMEVLHLRNVSFED
swissprot|FGR1_RAT|          IEVNGSKIGPDNLPYDQILKTAGVNTTDKEMEVLHLRNVSFED
swissprot|CEK2_CHICK|        VEVNGSKYGPDGTPYVTVLKTAGVNTTDKELEILYLRNVTFED
swissprot|FGR4_HUMAN|        IVINGSSFGADGFPYVQVLKTADINSS--EVEVLYLRNVSAED
swissprot|FGR4_MOUSE|        VVINGSSFGADGFPYVQVLKTTDINIS--EVQVLYLRNVSAED

pFGFR2                       AGNSIGISFHSAWLTVLPAPGRE-KEITASPDYLEIAIYCIGV
swissprot|FGR2_HUMAN|        AGNSIGISFHSAWLTVLPAPGRE-KEITASPDYLEIAIYCIGV
```

FIGURE 6.4: ClustalW Output

scale projects, a free license may be obtained, and scripts will help in the bulk production of models, in a similar manner to HMMer.

6.3.1 Inherent Differences between SAM and HMMer

SAM models include nine transitions, not just seven as with HMMer. The insert-delete and delete-insert transitions are included in the SAM models but not in the HMMer models.

Perhaps on a more esoteric note, the weighting of the input sequences is different between the two systems. HMMer places a higher weight upon the sequences that are used to build the model, whereas SAM places a much higher weight on the priors.

One other difference between the two systems is the usage of the null model. A basic null model would compare the difference between the sequence and the average frequency of amino acids in protein sequences. This would have problems when the sequence was highly biased towards some particular amino acid. So both SAM and HMMer have improved secondary null models for this scenario. HMMer bases the second null model on the average amino acid

```
HMMER2.0  [2.3.2]
NAME  P53
ACC   PF00870.9
DESC  P53 DNA-binding domain
LENG  196
ALPH  Amino
RF    no
CS    no
MAP   yes
COM   hmmbuild -f -F HMM_fs.ann SEED.ann
COM   hmmcalibrate --seed 0 HMM_fs.ann
NSEQ  7
DATE  Fri Apr 27 19:38:49 2007
CKSUM 2675
GA    14.6 14.6
TC    19.2 19.2
NC    10.2 10.2
XT       -8455      -4  -1000  -1000  -8455      -4  -8455      -4
NULT      -4  -8455
NULE     595  -1558     85    338   -294    453  -1158    197    249    902  -1085   -142
EVD  -10.738805   0.649352
HMM        A      C      D      E      F      G      H      I      K      L      M      N
         m->m   m->i   m->d   i->m   i->i   d->m   d->d   b->m   m->e
         -415      *  -2000
     1      1   2138  -1558  -1151  -1743   -986  -1022  -1362   -993  -1627   -850   -880
     -   -149   -500    233     43   -381    399    106   -626    210   -466   -720    275
     -    -25  -6676  -7718   -894  -1115   -701  -1378  -1415  -8607
     2    783   -593  -1423   -977  -1405  -1103   -842   -886   -827  -1257   -527   -815
     -   -149   -500    233     43   -381    399    106   -626    210   -466   -720    275
     -    -25  -6676  -7718   -894  -1115   -701  -1378  -9022  -8604
     3  -1376   -984  -3779  -3385  -1309  -3473  -2991  -3189   1942   -238   -149  -3135
     -   -149   -500    233     43   -381    399    106   -626    210   -466   -720    275
     -    -25  -6676  -7718   -894  -1115   -701  -1378  -9022  -8600
     4  -2089  -2190  -2567  -2787  -3372  -2253  -2663  -3607  -2895  -3590  -3215  -2612
     -   -149   -500    233     43   -381    399    106   -626    210   -466   -720    275
     -    -25  -6676  -7718   -894  -1115   -701  -1378  -9022  -8596
     5   -236  -1060   -462    640  -1429  -1251   -215  -1000     45  -1230   -454   -218
     -   -149   -500    233     43   -381    399    106   -626    210   -466   -720    275
     -    -25  -6676  -7718   -894  -1115   -701  -1378  -9022  -8592
     6   -452  -1565    -77    183  -1996  -1269   -195  -1690    184  -1758   -940   1848
     -   -149   -500    233     43   -381    399    106   -626    210   -466   -720    275
     -    -25  -6676  -7710   -894  -1115   -701  -1370  -9022  -8588
```

FIGURE 6.5: A Hidden Markov Model

composition of all the emission probabilities of the model. SAM simply uses the reverse of the submitted sequence for the null model score.

SAM increases the diversity of sequences by removing those that are more than 80% similar to any other sequence. This ensures that entropy is maximized, and that no single group of sequences is overrepresented.

The Target2k script is used to create an alignment from a seed and a database, as in the following example:

target2k -seed myseq -i pdb -out mydata

Target2k is optimized for alignments at the superfamily level. For models that are designed for the family level, the -family option may be used, and several other options may be 'tweaked' to achieve the proper level of detail.

Oftentimes a structural alignment is available to begin the process. In this case, the alignment may be used as a starting point rather than a single sequence. The option in this case is simply -seed, and the resulting model will frequently be superior to those beginning with the more common sequence. Of course, it may be that the alignment produced by the sequence will be just as good as the structural alignment, but in most cases the hand-curated structure will provide the best results.

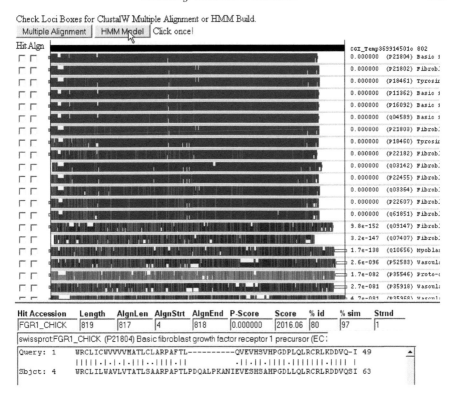

FIGURE 6.6: Graphic Output from the Second Iteration

If the structural alignment is adequate, we may of course bypass the Target2k step altogether. In this case, we may simply progress to w0.5 or one of the other model building scripts in an effort to eliminate the possibility of entering any false positives or incorrect alignments into the mix of data from which we will build our models.

The SAM package includes a program called 'modelfromalign' to build the alignment into a model, and another called 'buildmodel' that is used to refine the HMM. These are roughly the equivalent of the HMMbuild program in the HMMer package.

Many parameters and programs in the SAM package may lead one to believe that the system is too difficult to undertake. Fortunately, several scripts are provided with standard parameters that will provide excellent results with a minimum of trouble.

The w0.5 model building script is an excellent all-around program for making models that are designed to find remote homologs. The alignment is trimmed to remove sequences with greater than 80% homology, and the target entropy weighting is set at 0.5 bit per column.

Using the w0.5 script (or any of the scripts that are mentioned below) reduces the complexity and concern over the correct and reasonable use of parameters that are not part of the everyday lexicon of most bioinformaticists. The script to create a model from an alignment is reduced to

w0.5 mydata.a2m mydata.mod

The resulting model is somewhat similar to a HMMer model in that it has information presented for each position presented sequentially. The SAM model is perhaps less readable by humans because it does not explicitly state what each of the values is for. An example of the top part of a model is given below.

MODEL – Model from alignment file
/var/tmp/tmp-build-weighted-model-moai13.kilokluster.ucsc.edu-
25319/tmp.a2m
alphabet protein
FREQAVE
0.000000 0.000000 0.000000
0.000000 0.000000 0.000000
0.000000 0.000000 0.000000
0.085614 0.011725 0.044840 0.063195 0.042610
0.055045 0.027595 0.062151 0.047308 0.115600
0.027580 0.036079 0.043404 0.042919 0.059575
0.062503 0.056974 0.073352 0.011196 0.030736
0.085614 0.011725 0.044840 0.063195 0.042610
0.055045 0.027595 0.062151 0.047308 0.115600
0.027580 0.036079 0.043404 0.042919 0.059575
0.062503 0.056974 0.073352 0.011196 0.030736
Begin
0.000000 0.000000 0.000000
0.000000 0.000000 0.000000
0.000000 0.145949 0.657601
0.000000 0.000000 0.000000 0.000000 0.000000
0.000000 0.000000 0.000000 0.000000 0.000000
0.000000 0.000000 0.000000 0.000000 0.000000
0.000000 0.000000 0.000000 0.000000 0.000000
0.077850 0.016638 0.057323 0.065442 0.039839
0.062176 0.024317 0.060475 0.064413 0.082120

```
0.024556 0.047301 0.040289 0.041956 0.048788
0.067946 0.059891 0.072033 0.011738 0.034911
 1
0.000000 0.016698 0.181378
0.000000 0.837353 0.161021
0.025588 0.201579 0.470535
0.001362 0.000263 0.000974 0.001029 0.002171
0.000402 0.000334 0.002033 0.000733 0.005020
0.977935 0.000676 0.000187 0.000882 0.001014
0.000759 0.000756 0.001821 0.000422 0.001225
0.077848 0.016637 0.057322 0.065442 0.039839
0.062169 0.024326 0.060476 0.064420 0.082104
0.024543 0.047328 0.040276 0.041959 0.048780
0.067955 0.059895 0.072032 0.011739 0.034909
 2
0.527356 0.396879 0.071319
0.447055 0.401542 0.458146
 ...
```

The w0.7 script uses the same regularizer and prior set as the w0.5, but uses an entropy weighting of 0.7 bit, yielding a higher weighting on the sequences used, and less weighting on the recode3.20comp prior. Slightly closer homologs are allowed, through the use of a slightly looser trimming scheme of 85% instead of 80%. The w0.7 script is used when more specific models are desired, which may be useful in searching more closely related species.

Sequence logos provide a means of graphically viewing models that can quickly communicate the important features of a protein family to a researcher. For this type of view, we want higher weighting on the sequence data, and less on the priors. The w1.0 script will build models that more clearly show the details of the logo. Trimming is performed at the 90% level, allowing more sequences and more redundancy into the model build. While the w1.0 script, which uses a 1 bit per column entropy weighting, is primarily used for building logos with the makelogo program, the models may certainly be used for searching. The results will typically even be more sensitive than models built with w0.5, although they will likely be less specific. That is, they may find a greater number of true positives, but will also find a greater number of false positives.

Once a model is created, it can be used to make highly accurate alignments. To align a set of sequences to an existing model, the align2model program is used. The syntax follows the following pattern:

align2model newalign -i mydata.mod -db new.seq

This will create an alignment file called newalign.a2m from the alignment of the new.seq data to the mydata.mod model.

The raw alignment is functional, but not very human readable. The prettyalign program will take the alignment created by the align2model program and clean it up for publication or further use. The command in this case would be

prettyalign newalign.a2m -l80 > newalign.pretty

The l80 (lowercase L 80)option forces the output to contain 80 characters per line.

The model may be compared to a database of sequences through the use of the hmmscore program, which is similar to the hmmsearch program in the HMMer package. The syntax is as follows:

hmmscore example – mydata.mod -db new.seq -sw 2

The name of the run is 'example', the model is mydata.mod, the database sequence file is new.seq and the -sw 2 option means that the search will be in local mode.

The models are only as good as the alignments that they are built upon, and SAM provides two useful tools to modify the a2m alignments. One is called select-by-key-residues and the other is select-by-gapless-regions. Both of these programs delete lines in an a2m alignment that do not meet certain criteria.

The script called select-by-key-residues takes a description file of key residues that must be present, and removes all sequences from the alignment that do not include these residues. The residue specified at a location may be a gap, and if two or more residues are permitted at a location, they may both be specified. A residue file like

IL120 H-166 S240

therefore means that the alignments must have an isoleucine or a leucine at column 120, a histidine of a gap at column 166 and a serine at column 240.

Another method of restricting the alignments is with the -nogap option. If you want to force an alignment to have no gaps in the the area of residues 25 through 40, and in another region between 70 and 80, you may create a file called something like 'nogap.txt' which has these regions spelled out one per line, with a space between them:

2540

7080

Then we run the program select-by-gapless-regions in this way:

select-by-gapless-regions -nogap nogap.txt > myseq.a2m >
mynewseq.a2m

This command will delete any lines in myseq.a2m that have gaps in either of these two regions and save the results in mynewseq.a2m.

Note that ClustalW also has a similar feature. ClustalW follows the same logic, but it works somewhat differently, by forcing an alignment to be 'pinned' around a certain residue or group of residues before the alignment begins.

Building alignments with SAM can have the unintended effect of expanding the sequences to include members of another family or superfamily. When this happens, the alignments tend to expand rapidly with the contaminating data, which then grab additional members of the unwanted family with successive iterations. To protect against this occurrence, the -all option will provide the output from each iteration with an addition to the output name, so that mydata.seq will yield

mydata_1.a2m

mydata_2.a2m

mydata_3.a2m

...

and these output files may be examined to see when the interfering sequences began to be incorporated. This problem is also endemic to other automated systems such as PSI-BLAST.

As with HMMer, SAM searches may be local/local local/global or global/global. The option here is the SW variable. SW 0 refers to a fully global search, that is, all of the sequence matches all of the model. SW 1 is perhaps the most commonly used, with part of the sequence matching all of the model (referred to as glocal). SW 2 is local with respect to the sequence and to the model.

SAM has some advantages over HMMer in the area of multidomain proteins. Free-insertion modules (FIMs) remove the match states in between domains, providing a good way of connecting complex proteins. An even better method, however, is to trim out the domains and use them separately.

Once a large number of models have been generated, they may be concatenated into a large library and scored against a sequence database as one large job. SAM tends to be even slower than HMMer, though, so the resulting searches can take an inordinately long time to complete. An accelerator system called kestrel has been developed to speed these searches, but no commercial accelerator is available.

6.4   BUILDING PSSM DATABASES FOR RPS-BLAST

Building a database of PSSMs (Position Specific Scoring Matrices) is generally considered to be faster and easier than constructing an equivalent collection of HMMs. Certainly the resulting searches are much faster than running HMMPfam or SAM! NCBI also provides a dynamic programming method for searching the databases. This method, known as Impala, has become much less popular since the release of RPS-BLAST (Reverse Position Specific BLAST).

PSI-BLAST begins with a single sequence, finds matches using an initial BLAST search, aligns those matches, builds a PSSM or PSI-BLAST profile as it is also called, and then searches the database again with that profile. Any new hits are used to improve the profile, and the process is iterated until no new sequences are found.

The profiles that are generated in this process may be saved for further use. Another program called RPS-BLAST (Reverse Position Specific BLAST) can be used to search this profile against any database that you like, and a collection of these profiles can provide a quick way to analyze complete genomes.

Despite the potential loss in sensitivity that you may experience from using PSSMs instead of HMMs, this will be the best choice for many applications. Building your database using PSI-BLAST should be considered by anyone who is beginning this type of project.

Start by running PSI-BLAST with your example sequences against an appropriate database using the –C option. This will save the scoring profiles with a .chk extension. We will use the example of running a collection of peanut protein sequences against the nonredundant uniprot database.

Blastpgp -d uniref90 -i peanut1.faa -j 5 -t T -s T -C peanut1.chk

Blastpgp -d uniref90 -i peanut2.faa -j 5 -t T -s T -C peanut2.chk

...

The above command will BLAST each peanut.faa amino acid sequence against the Uniprot 90 database. Some of the options in this line may not be familiar, so we whould take a closer look at what they are, and how they may help to produce better results.

The -J command is essential in that it sets the number of iterations that PSI-BLAST will make at a maximum against the database. By default, this option is set to one, which is the same as a standard BLASTP search. After each iteration, the results are compared to the time before, and if no new hits are found, the process is halted. So, the -j value is only reached if the process does not converge on its own.

We will set -t to true, in order to get the best profile. This option adjusts the score based on composition. If there are any composition biases in the query and target sequences, the PSSM can get corrupted and wind up finding false hits to any other sequence with that same bias. The -t parameter avoids this problem by introducing some low-entropy false positives to the alignment.

The -s option is also set to True, in order to ensure the highest quality in the initial seed alignments. This will force PSI-BLAST to use true Smith-Waterman based alignments which is likely to slow things down somewhat, but can make a great deal of difference to the end result.

Makemat is then run on these binary profiles to convert them to a portable ASCII format. Some unusual requirements are found in makemat, but these requirements are simply because this system is designed to create an entire database at once. We need to have all the sequences associated with these profiles named with an extension that starts with c, so we have to rename the files from peanutX.faa to peanutX.caa.

Next we need two files, one with a list of the profile names and the other with a list of the related sequence names. The former will have an extension of .pn (profile name) and the latter will have an extension of .sn (sequence name). The order in both files must be the same, as shown below.

Test.sn	Test.pn
Peanut1.caa	Peanut1.chk
Peanut2.caa	Peanut2.chk
Peanut3.caa	Peanut3.chk
...	...

The makemat program is run to convert the checkpoint files into PSSMs.

Makemat -P test The output from the makemat program consists of peanut1.mtx, peanut2.mtx, etc., in this example. A file consisting of the names of all the portable ASCII matrices is created with a name of test.mn in this example. Auxiliary information is stored in a file called test.aux.

Copymat is then run as a secondary preprocessor, creating a large .rps file. Be sure you have plenty of RAM for this process! Then the profiles must be formatted with formatdb before they are accessible with RPS-BLAST.

Copymat -P test -r T Then the BLAST database of all the 'master sequences' of the profiles must be formatted, as in the following example:

formatrpsdb -t test -i test.pn -o T -f 9.82 -n test -S 100.0 When the parameter '-f' is supplied with formatrpsdb, the word score threshold for detecting and extending hits in RPS-Blast, will determine the size of the search database. A lower threshold will result in larger databases and slightly increased search sensitivity, at the cost of additional memory requirements and reduced search speed.

Matrices distributed for creating RPS-Blast search databases are scaled by parameter -S, using a factor of 100 in this example. Setting the -o value to true allows the database to be indexed, just as in formatdb.

While the ASCII text formatted matrices are portable between different computer architectures, the binary matrices are not. Therefore, if you need to move your files between Windows and Solaris systems, you will need to move the ASCII files and then format them on the systems where they will be utilized.

The RPS-BLAST executable is run in the same way and with similar options as a standard blastall search:

RPSBLAST -i myseq.faa -d test.rps -o myseqtest.out -e 0.001

The obvious advantage of building a database in RPS-BLAST format is that the speed is dramatically greater than any other Hidden Markov Model or profile method (except those running on special accelerator hardware). A more obscure reason may be that the targets are more portable in the ASCII version once built, that is, a lot more people have BLAST running on their computers than HMMer or SAM. The setup of BLAST is extremely simple, and no license is required. The loss of sensitivity is not so great for most purposes that projects cannot be completed.

Please note that SAM does a much better job of building models than either HMMer or PSI-BLAST. If you have the time and the ability, and if it is important to you to have models with the highest quality, the preferred method would be to build the models with SAM and then convert to an RPS-BLAST style database using the sam2psi program. The precise amount of improvement depends greatly upon the data utilized and the project. After all, in many cases the top hit is all that is of interest, and precise modeling is not necessary. For others, however, the highest precision and sensitivity are essential.

6.5   BUILDING REGULAR EXPRESSION DATABASES

Tools for finding and building regular expression type databases include the well-established PRATT2 system, which is used to build the ProSite database, PatMatch from the Arabidopsis group at TAIR, and the eMATRIX/ eMOTIF system from the Brutlag lab at Stanford. NCBI has an option to the BLASTPGP program called PHI-BLAST that accepts prosite-style records. For nucleic acid motif searching there are algorithms such as smartfinder and SMOTIF.

The standard method for motif searching is to compare a sequence to a database of signatures, or a signature against a database of sequences. The program QuasiMotiFinder extends the capability to include searches starting with MSAs. The authors of QuasiMotiFinder have discovered that this approach reduces both the false positive rate and the false negative rate.

The PRINTS database uses multiple motifs, which provide far fewer false positives than single motif systems. The database may be searched via a BLAST search for matches to sequences that are matched in the database, or searches can be made against fingerprints contained in the database directly through the use of FingerPRINTScan. This method is notably more specific than the BLAST method, and can make the hierarchies of the related families plain.

While the PRINTS database is derived through an entirely manual process, an automated supplement known as prePRINTS is also available. The pipeline to automatically devise this subsidiary database takes protein clusters from ProDom. A MSA is made with ClustalW or DIALIGN. Motifs derived from the MSAs are compared in an iterative manner to UniProt. Annotations for the fingerprints in prePRINTS are derived automatically from SwissProt.

ProDom itself is made from the Uniprot database with the use of PSI-BLAST and a program called MKDOM2. The MKDOM2 process begins by removing low complexity data with the SEG filter. Without this step, you would have sequences clustered together even though they had nothing in common except for a string of repetitive characters.

Next, the shorter sequences are removed—typically the minimum is set at 20 characters. The shortest sequence that remains will be chosen as the first to be tested. This sequence is BLASTed against itself to look for repeats. If no repeats are found, then the sequence will be used in its entirety. If a repeat is found, then it will be used as the seed.

The query sequence is compared to the rest of the database using PSI-BLAST. Any segments are considered at this point to be family domains. Overlapping domains are split if the overlap is not long enough—this is set at 10 residues for a default. The process is iterated through the entire database until no sequences remain.

An accompanying program called XDOM is provided to allow viewing of the alignments. The domain structure of each protein is visualized with different hashing patterns. This enables the user to examine the structure and can usually point out the most obvious difficulties. For example, if domain A is always found just upstream of domain B, then these may not be two domains but one that was incorrectly split.

Blocks are defined as MSAs of highly conserved protein regions. These blocks are automatically generated from the most highly conserved protein groups in the Prosite database. Calibration against the SwissProt database provides a means to evaluate the probabilities of a chance hit. The alignments in BLOCKS+ and PRINTS are used to generate the motifs in the eMOTIF database.

6.6 SUMMARY

The development of a custom database may be a brief exercise for a single specific purpose, or it may be an ongoing project that will last for years. The usability of that database may be extremely broad, or it may be of interest only to you. The direction that you take in this task depends entirely upon the goals that you outline at the beginning of the project.

Of the Profile-HMM databases, most will be built in HMMer or SAM format. Each has their proponents, and the decision will be left to the reader, as there is no 'best' system. In general, SAM is considered to have the superior method for generating models, so one possibility that must be considered is to generate all the models in SAM format, and then convert the resulting database to HMMer or PSI-BLAST format, depending on your preferences.

Remember too that not all data analysis systems respond in the same fashion. One person's data may be well represented in a method that is entirely inappropriate for another. I heartily recommend that a small subset of the project be undertaken, and several methods tried before deciding a priori that one method is superior. For example, the data might be represented first as a set of regular expression motifs, then as an RPS-BLAST database, and so forth. Evaluations may then be used to guide the full project.

6.7 QUESTIONS

1. Write out the pattern for the 14-3-3 protein signature in both Prosite and Perl formats.

2. Extract a protein sequence from the Uniprot database as guided by your instructor. Build it into a Hidden Markov Model using the SAM-T06 website.

3. Convert the Hidden Markov Model from problem 2 to HMMer format using the convert.pl script.

4. What percentage of SwissProt proteins are represented in ProDom?

5. What is the relationship of the BLOCKS database to BLOCKS+?

6. How is Pfam related to ProDom?

7. Build a small RPS-BLAST database from a fasta dataset provided by your instructor and provide search results of this database against the Uniprot-SP database.

References

[1] http://www.ncbi.nlm.nih.gov/sites/entrez?Db=PubMedCmd
=ShowDetailViewTerr

[2] M. Wistrand and E. Sonnhammer, "Improved Profile HMM Performance by Assessment of Critical Algorithmic Features in SAM and HMMER," *BMC Bioinformatics* **6**, 99 (2005).

[3] M. Wistrand and E. Sonnhammer, "Transition Priors for Protein Hidden Markov Models: an Empirical Study Towards Maximum Discrimination," *J. Comput. Biol.* **11**(1), 181–93 (2004).

[4] R. Hughey and A. Krogh, "Hidden Markov Models for Sequence Analysis: Extension and Analysis of the Basic Method," *CABIOS* **12**(2), 95107 (1996).

[5] K. Karplus, C. Barrett, and R. Hughey, "Hidden Markov Models for Detecting Remote Protein Homologies," *Bioinformatics* **14**(10), 846856 (1998).

[6] A. Krogh, M. Brown, I. S. Mian, K. Sjolander, and D. Haussler, "Hidden Markov Models in Computational Biology: Applications to Protein Modeling," *Journal of Molecular Biology* **235**:15011531 (1994).

[7] K. Karplus, R. Karchin, C. Barrett, S. Tu, M. Cline, M. Diekhans, L. Grate, J. Casper, and R. Hughey, "What is the Value Added by Human Intervention in Protein Structure Prediction?" *Proteins: Structure, Function, and Genetics* (2001).

[8] R. Wheeler and R. Hughey, "Optimizing Reduced Space Sequence Analysis," *Bioinformatics* **16**(12), 10821090 (2000).

[9] K. Karplus, C. Barrett, M. Cline, M. Diekhans, L. Grate, R. Hughey, "Predicting Protein Structure using Only Sequence Information," *Proteins: Structure, Function, and Genetics*, Supplement 3, 121125 (1999).

[10] J. Park, K. Karplus, C. Barrett, R. Hughey, D. Haussler, T. Hubbard, and C. Chothia, "Sequence Comparisons Using Multiple Sequences Detect Twice as many Remote Homologues as Pairwise Methods," *Journal of Molecular Biology* **284**(4), 12011210 (1998).

[11] R. Karchin and R. Hughey, "Weighting Hidden Markov Models for Maximum Discrimination," *Bioinformatics* **14**(9), 772782 (1998).

[12] C. Tarnas and R. Hughey, "Reduced Space Hidden Markov Model Training," *Bioinformatics* **14**(5), 401406 (1998).

[13] K. Karplus, K. Sjolander, C. Barrett, M. Cline, D. Haussler, R. Hughey, L. Holm, and C. Sander, "Predicting Protein Structure Using Hidden Markov Models," *Proteins: Structure, Function, and Genetics*, Supplement 1 (1997).

[14] K. Sjolander, K. Karplus, M. Brown, R. Hughey, A. Krogh, I. S. Mian, and D. Haussler, "Dirichlet Mixtures: A Method for Improving Detection of Weak but Signicant Protein Sequence Homology," *CABIOS* **12**(4), 327345 (1996).

[15] C. Barrett, R. Hughey, and K. Karplus, "Scoring Hidden Markov Models," *CABIOS* **13**(2), 191–199 (1997).

[16] J. A. Grice, R. Hughey, and D. Speck, "Reduced Space Sequence Alignment," *CABIOS* **13**(1), 4553 (1997).

[17] A. Krogh, I. S. Mian, and D. Haussler, "A Hidden Markov Model that finds Genes in E. coli," *Nucleic Acids Research* (1994).

[18] R. Hughey and A. Krogh, "SAM: Sequence Alignment and Modeling Software System," Technical Report UCSC-CRL-96-22, University of California, Santa Cruz, CA, 1996.

[19] K. Karplus, "Regularizers for Estimating Distributions of Amino Acids from Small Samples," Technical Report UCSC-CRL-95-11, University of California, Santa Cruz, CA, 1995.

[20] C.-M. Hsu, C.-Y. Chen, and B.-J. Liu, "MAGIIC-PRO: Detecting Functional Signatures by Efficient Discovery of Long Patterns in Protein Sequences," *Nucleic Acids Res.* **34**(web server issue), W356–W361 (2006).

[21] R. Gutman, C. Berezin, R. Wollman, Y. Rosenberg, and N. Ben-Tal, "QuasiMotiFinder: Protein Annotation by Searching for Evolutionarily Conserved Motif-Like Patterns," *Nucleic Acids Res.* **33**(suppl_2), W255–W261, (2005).

[22] I. Jonassen, J.F. Collins and D. Higgins, "Finding Flexible Patterns in Unaligned Protein Sequences," *Protein Science* **4**(8), 1587–1595 (1995).

[23] I. Jonassen, "Efficient Discovery of Conserved Patterns Using a Pattern Graph," *Comput. Appl. Biosci.* **13**(5), 509–522 (1997).

[24] P. Puntervoll, R. Linding, C. Gemnd, S. Chabanis-Davidson, M. Mattingsdal, S. Cameron, D. Martin, G. Ausiello, B., Brannetti, A. Costantini, F. Ferr, V. Maselli, A. Via, G. Cesareni, F. Diella, G. Superti-Furga, L. Wyrwicz, C. Ramu, C. McGuigan, R. Gudavalli, I. Letunic, P. Bork, L. Rychlewski, B. Kster, M. Helmer-Citterich, W. Hunter, R. Aasland, and T. Gibson, "ELM Server: a New Resource for Investigating Short Functional Sites in Modular Eukaryotic Proteins," *Nucleic Acids Res.* **31**, 3625–3630 (2003).

[25] J. Huang and D. Brutlag, "The eMOTIF Database," *Nucleic Acids Res.* **29**, 202–204 (2001).

[26] J. Henikoff, E. Greene, S. Pietrokovski, and S. Henikoff, "Increased Coverage of Protein Families with the Blocks Database Servers," *Nucl. Acids Res.* **28**, 228–230 (2000).

[27] C. Nevill-Manning, T. Wu, and D. Brutlag, "Highly Specific Protein Sequence Motifs for Genome Analysis," *Proc. Natl. Acad. Sci. USA* **95**, 5865–5871 (1998).

[28] S. Henikoff, J.G. Henikoff, and S. Pietrokovski, "Blocks+: A Nonredundant Database of Protein Alignment Blocks Derived from Multiple Compilations," *Bioinformatics* **15**(6), 471–479 (1999).

[29] E. de Castro, C. Sigrist, A. Gattiker, V. Bulliard, P. Langendijk-Genevaux, E. Gasteiger, A. Bairoch, and N. Hulo "ScanProsite: Detection of PROSITE Signature Matches and ProRule-Associated Functional and Structural Residues in Proteins," *Nucleic Acids Res.* 34(web server issue), W362–W365 (2006).

Chapter 7

Speeding Your Searches

7.1 INTRODUCTION

Chapter 7 describes methods for speeding up searches. An HMMPfam search can be slow with a large batch of proteins, but now that the reader is aware of additional databases, the problem is compounded dramatically. Similarly, the researcher with a group of HMMs has a greater number of proteins to search against, and the number of proteins is growing faster than the progress in processor speed.

Fortunately, many solutions exist to help overcome this problem, and the goal of this chapter is to help the reader select the right one. Server farms had a tremendous rise in popularity due to the tremendous cost efficiency of common off-the-shelf technology. Unfortunately, the use of these systems has not been able to keep up with the ideas of researchers. Gollerys' first law of biocomputing states that the amount of computing power available will always be exceeded by the ability of a researcher to create new questions. In other words, it is much simpler to devise an experiment that requires 100 times the amount of compute power than it is to enlarge the size of the cluster by 100 times.

In the analysis of biological data, two time values are important. The CPU time is most oftentimes quoted and discussed in benchmarks. This is certainly important, as the ability to obtain output in a reasonable timeframe is paramount when competitors are working on a similar problem or a deadline is imminent.

The time required of the bioinformaticists to accomplish a task is less discussed but often more important than CPU time. Splitting a job into a thousand pieces for distribution across a grid will get that job done faster but involves more work beforehand as well as afterwards.

Another consideration for biocomputing solutions is energy usage to accomplish your goals. Processors generate heat, and a large number of processors will generate a large amount of heat. Furthermore, every kilowatt that is utilized in a server farm must be removed by a greater amount of power in air conditioning. Large computing facilities can easily spend hundreds of thousands of dollars in cooling costs alone.

Finally, the cost to rent floorspace should be considered as a part of the e-quation. The most efficient solution in a 'dollar-per-teraflop' calculation may involve more square feet of floorspace than are currently available. Once when working at a small university, my server room was completely full, and we could not expand the cluster until a new building was constructed. Then when the new facility was designed, no thought was given to informatics, insufficient server space was planned, and the expansion had to be canceled.

This concept of space efficiency as an essential component of system planning is behind the drive towards blade servers. While the cost per CPU is higher with a blade system than with standard clusters, the density of processors per square foot is much higher.

In short, all of the relevant factors must be considered when designing a solution to achieve the greatest benefit on a long-term basis, and the cost to own a system must be a factor as well as the cost to obtain it.

7.2 PICK YOUR TARGETS CAREFULLY

Once you have done much analysis with HMMs, you will come to the realization that analyzing a large amount of data with SAM or HMMer can tax even the largest server farm. Sometimes the simplest solution is the best, and so we will look at the most trivial changes first. Only when those possibilities are exhausted should you consider bringing out the more powerful tools.

7.2.1 Target Triage

Let us imagine that we have set of 10,000 proteins from a group of organisms that we are studying. Running these sequences on a single CPU against the Pfam database can easily take a month, depending upon the length of the sequences, the speed of the CPU, and so forth. If you are looking for a particular class of protein HMMs, for example, protein kinases, consider simply extracting them out of the Pfam database, to make a smaller target.

Performing target triage takes more upfront thought and effort. Procedural mistakes are common; missing some of the domains of interest when the database is constructed means that either the entire job has to be rerun or the missing models need to be run separately and the results concatenated.

Perhaps more commonly, the goals of the project are changed during the review of the initial data. The intrepid bioinformaticist analyzes his 10,000 proteins and identifies the kinases in the set, and then the collaborators decide that they want to include a study of transferases! So, the process needs to be redone.

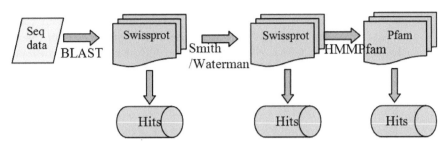

FIGURE 7.1: A Simple Data Distillation Column

7.2.2 Data Distillation Column

If the idea of target triage is not applicable to your project, then you might be able to use what is sometimes referred to as a 'data distillation column'. This involves running a prefilter to make a first pass at the data. With the bulk of the data weeded out, only a relatively small subset is submitted to the HMMPfam search. This may work because the prefilter is sufficient to annotate the data or because some rules are able to eliminate a portion of the sequences.

In a typical data distillation column, as seen in figure 7.1, BLAST or some other fast, heuristic algorithm is used for a first step. The output from this step may be separated into sequences that have matches in the database (hits) and those that do not (nohits). Since the Smith-Waterman algorithm is more sensitive, yet slower than BLAST, the nohits data can be rerun with Smith-Waterman to find a few more hits and thus pare down the data considerably. Only the data which still have not found a hit in the previous two searches will be run against Pfam.

The amount of data reduction that you will achieve depends not only on the data but also on the databases chosen and the parameters used to decide between the 'hits' and the 'nohits.' The Max Planck Institute has used a variation on this technique in their 'FastHMMer' system, for a reported speedup of 30 times over a straight HMMer search. Further extension of the column to other HMM databases or other algorithms may provide additional information for the most difficult sequences to analyze.

7.3 FORMAT SELECTION

RPS-BLAST is many times faster than HMMer or SAM, although it is a noticeably less sensitive algorithm. If you don't mind that this may change your results somewhat, you can generate your targets in RPS-BLAST format as outlined in chapter 6. If you are interested in Pfam instead of generating your own databases, remember that using the Conserved Domain Database can also speed your search. The CDD is more than just Pfam, as it also includes data from COG, KOG, Smart and CD, which is a database from conserved domains at the NCBI. The components may be utilized separately or all together, formatting them as follows:

formatrpsdb -t SMART.v4.0 -i Smart.pn -o T -f 9.82 -n Smart -S 100.0

formatrpsdb -t Pfam.v.11.0 -i Pfam.pn -o T -f 9.82 -n Pfam -S 100.0

formatrpsdb -t COG.v.1.0 -i Cog.pn -o T -f 9.82 -n Cog -S 100.0

formatrpsdb -t KOG.v.1.0 -i Kog.pn -o T -f 9.82 -n Kog -S 100.0

formatrpsdb -t CDD.v.2.03 -i Cdd.pn -o T -f 9.82 -n Cdd -S 100.0

These databases may be run all at once for the full CDD analysis as given on the NCBI website, or they may be combined in any group of subsets for a custom search.

7.4 OPTIMIZED SOLUTIONS

Ultimately, you probably have to get down to running a job. Several optimized solutions exist, each with its own costs and benefits. To decide which solution provides the optimal answer for your project, lab or facility, you must weigh the amount of resources that are available with the utilization of those resources. Several optimized solutions for HMMer are available, with different techniques to achieve their higher speeds.

First make sure that you are running HMMer in the most efficient manner. For many searches (particularly short searches), disk I/O is a major component of the runtime. This may be reduced significantly by converting the database into binary form. The program to use is the HMMconvert module of the HMMer package:

Hmmconvert -b Pfam Pfam_bin

This method will reduce the size of the database by approximately 50%, and the retrieval time will be shortened proportionally.

Another check that needs to be made is that the database should be on a local hard drive. Mounting disks remotely is now so ubiquitous that it is sometimes not thought of as an issue anymore. For a word processing document this may be true, but for a database that is several hundred megabytes, it is not. The full impact depends on the size of the database and the speed of the network, as well as the number of queries that are submitted.

Using all the available cores is also crucial to getting the most out of your server. Check the efficiency of the system with the 'top' command, or some other method to check system performance. The option to specify the number of CPU's or cores is simply –cpu, as in the following command:

HmmPfam –cpu 4 Pfam test.seq

This command will run the test.seq file against Pfam using four CPUs, or the equivalent, such as a dual CPU system with dual core CPUs. The program is supposed to use all available CPUs as a default, but in practice it doesn't always happen unless the number is explicitly specified.

7.4.1   MPI-HMMer

If the searches are still taking too long, you may want to distribute the search over a cluster of computers. MPI-HMMer is an optimized version of hmmsearch and hmmPfam. The older PVM version of the original 2.3.2 HMMer code was refactored into MPI code, but retained the options and functionality of the original program. The code was hand optimized for speed and a number of improvements were made. As a result, the program is reported to be about 12 times faster than the 'off-the-shelf' code.

7.4.2   SledgeHMMer

Some researchers at the San Diego Supercomputer Center have made some interesting optimizations in a package they call Sledge HMMer. SledgeHMMer uses the UNIX file locking scheme rather than MPI or PVM to run the distribution of the search across several nodes. In this manner, compute nodes may join and leave the queue as they become available. This scheme requires that each node access a lock-file, which limits the operating system choices somewhat. On the other hand, load balancing is nearly perfect, even when processors of different speeds are added to the system.

The SledgeHMMer website also precalculates results for a large number of queries. In this way, any queries that match queries that have already been run can be rapidly returned. Precalculated queries are maintained in a hash database for rapid retrieval, so that CPU utilization is minimized.

SledgeHMMer has a third trick up its sleeve. The entire Pfam database is kept in memory so that it does not have to be read from the hard disk for each query. Reading the rather large Pfam database file from the hard disk is a major contributer to the runtime of the search, particularly for a single query. Moving the file into the much faster memory is a good way to achieve a speedup.

This last optimization may be used in other methods as well, if you have a bit of knowledge about your operating system. A virtual disk may be established in memory, to create what is sometimes called a 'RAM disk'. Retrieving data from a RAM disk is much faster than a hard disk system. This is an expensive way to store data in comparison to a hard disk, but it may improve performance considerably, and will likely be much less expensive than buying additional systems.

7.4.3   ClawHMMer

Graphical processing units are extremely powerful processors that have been developed with the demands of gaming aficionados in mind. Since these processors (frequently referred to as GPUs), are included in all desktop computers, it is a natural progression to put them to work in computational biology. Three researchers from Stanford University ported the Viterbi algorithm to run on GPUs via the Brook programming language. The results were quite impressive, with speeds up to about twenty times a standard CPU. The program, called ClawHMMer, is freely available on sourceforge.net.

ClawHMMer actually uses the GPU as part of an early reject process in hmmsearch, with high probability traceback calculations performed on the CPU. If the HMM is too large, the batch size must be reduced, making the system much less efficient. This traceback is essential to the NULL2 correction phase of the process. The developers therefore realized that the GPU would not be an effective platform for hmmPfam. Since most large HMM jobs are based on hmmPfam, ClawHMMer has limited use.

7.4.4   JackHMMer

As the number of cores on a server or desktop machine continues to grow, it is interesting to note that other classes of processors have long had multiple cores. The Intel IXP 2850 has 16 cores. Individually, these would not be competitive with running HMMer on an Intel or AMD CPU, but collectively they might be able to perform quite well. The JackHMMer project was a proof-of-concept system that did show that the IXP could run at 2–4 times the speed of a single core CPU.

7.4.5   Commercially Available Solutions

Southwest Parallel Software and Logical Depth both produce optimized versions of HMMer code that they sell commercially. One series of tests of the SPSPfam program from Southwest Parallel showed a speedup of 3–60 times over the standard HMMer code, with the most common speedup being around 10–20X.

Other companies that have produced optimized software versions of the HMMer package include Biocceleration, as part of the Gencore package, and Scalable Informatics. Biocceleration reports a speedup of roughly 10x, and Scalable Informatics reports a speedup of up to 2.5X.

Running code from any one of these companies on all nodes of a server farm will provide a very high throughput at a much lower price than upgrading the server farm. In addition to a lower acquisition cost, running these packages across a cluster will incur no additional charges for electricity, administration or cooling. These charges can easily surpass the initial acquisition costs for a cluster upgrade, and must be figured into the calculations when considering whether to put money into hardware or software.

7.5 ACCELERATED COMPUTING

Computational biology can be extremely important to the success of grant applications and ongoing research projects. Special-purpose computers, sometimes called accelerators, have been designed to handle large amounts of data with a performance advantage. Accelerators typically utilize a type of processor known as a Field Programmable gate Array, or FPGA for short. Companies have connected the FPGA chips to a board, which is then connected to the host computer through a PCI-e card slot or a USB cable.

An accelerator can help tremendously in the area of HMM searches. Some applications do not accelerate very well, but both hmmsearch and hmmPfam show tremendous benefits with acceleration. At the time of this writing, TimeLogic and Progeniq are the only accelerator makers that offer HMMer searches, but there will be others coming out soon.

The Kestrel project led by Richard Hughey was originally designed to accelerate the SAM searches, but has more recently been made more application unspecific—that is, Kestrel can now accelerate a wider range of applications than just SAM. One test that I have seen showed a speedup of 20–40x by the Kestrel board over a conventional CPU.

In addition, there are some open-source projects in the works. While FPGAs are notoriously difficult to program, a great deal of work is being done to alleviate this bottleneck. Mitrion-C, Handel-C and Impulse-C are designed to make the transition from software to hardware easier. Mitrion-C is notable in that they have given a special emphasis to the acceleration of bioinformatics algorithms. I expect that there will someday be multiple accelerated HMM search variants just as there are now multiple software optimized versions on the market.

FPGA acceleration allows for large jobs to be accomplished in a short period of time. Speedups are commonly in the range of 50 to 1000 times. As a result, a job that would take a week on a conventional CPU would complete in 10 minutes. The benefits of running at this speed versus the cost savings of using the optimized software or the original software are left to the reader. Even though accelerators are much less expensive than they were a few years ago, if running large jobs is rarely done, then an accelerator is not called for. On the other hand, as the amount of data continues to increase, the need for acceleration will also rise.

7.6 GENEWISE

GeneWise, although an excellent package of programs, can be unbearably slow. The author, Ewan Birney, has stated that the code is most definitely not optimized for speed, but also believes that optimizations will not be able to change much. Simply put, nucleic databases are very large, and the GeneWise program is very computationally intensive.

Instead of using the GeneWisedb executable to analyze large amounts of nucleic data against the Pfam database, there are two scripts that use BLAST as a sort of prefilter to lessen the computational burden:

Halfwise—a script that compares a DNA sequence to a protein database built with sequences that are derived from the Pfam full alignments, using BLASTX.

Blastwise—a Perl script that compares a DNA sequence to a protein database using BLASTX and then extracts the hits and runs GeneWise on those extracted sequences.

To run either of these scripts, you need to first install BioPerl and BLAST. To run Blastwise, some system for retrieving sequences from a database (such as SRS) must also be installed.

Actual execution of the scripts is quite trivial. Halfwise is run by a command with the following syntax:

> halfwise test.seq > test.seq.hlf
>
> And blastwise is quite similar:
>
> blastwise test.seq > test.seq.blw

ESTwise is equally difficult to use in a large-scale fashion, because the EST databases are large and the interest here is in sensitivity and the overall quality of the alignment. ESTwise is therefore typically run standalone, without the tricks used with the Halfwise and Blastwise scripts.

In the first place, there are multiple algorithms that make up ESTwise. The 3:33 algorithm is very sensitive, but considerably slower than the 3:12 algorithm. Run the 'quick' version (quick being a relative term here) through the following command:

> estwise -alg 312Q

The wise tools must be built with thread support to spread the search over multiprocessor/multicore systems. Use the -pthread option to build the support into the executable.

ESTs are frequently repetitive, sometimes to the extreme. Running ESTwise on the raw data is wasteful of resources. Clustering these data sets with

CAP3, or some other clustering and assembly tools, will improve the quality of the data as well as reducing the total volume.

Reducing the size of the target is also a possibility, as mentioned for the HMMer searches above in the 'Target Triage' section. If you are not interested in Domains of Unknown Function (DUFs), for example, the database size may be reduced by several hundred models by eliminating these entries, and the search time will be reduced proportionally.

An accelerator system such as the DeCypher can compare ESTs to the Pfam (or other HMM) database at a greatly increased rate, typically several hundred times faster than a single CPU core. No algorithmic tricks such as Blastwise or Halfwise are utilized (although this could be introduced at a later date to further increase the throughput).

RPS-BLAST can also be a solution for this situation, as it is for proteins. Simply use the -p F option to tell the program that the input is nucleotide instead of protein and the conceptual six frame translations will be generated from the ESTs and each of these will be run against the target. While running six jobs instead of one will slow down the search proportionally, it will still be dramatically faster than ESTwise.

7.7   SUMMARY

The analysis of large amounts of protein sequence data with HMMs can prove to be a daunting task due to the computational complexity of the algorithms that are involved. The number and variation of solutions that are available to solve this problem are an indication of the breadth of the problem. It is clear that the simple increase in CPU speed has not been sufficient to keep up with the influx of data.

Solving the speed issues will not be a simple solution that will be the same for every situation. Finding the optimal system will involve a careful analysis of the available tools. In this chapter we have started with an examination of the best ways to get the most performance out of the freely available code. The HMMer website has many executables built for a wide variety of architectures. Selecting the best one for your facility, running it with a local (not network mounted) database, and perhaps making use of a RAM disk should be the first priority.

Optimized software from commercial entities has a terrific advantage over simply buying more CPUs for a cluster. Frequently the acquisition cost is much lower, and the cost of ownership is much better. Whether this kind of solution would work well for a given situation is easily determined, as the vendors typically allow a trial period at no cost.

For those with a regular need for high-throughput HMM searches, the purchase of an accelerator or cluster of accelerators will provide the highest possible performance. Once these special-purpose systems were quite expensive, but now the price has come down considerably while the overall throughput has increased. If you find that your cluster is oversubscribed, and a large portion of the utilization comes from running HMM jobs, then you probably will find that adding an accelerator will be a less expensive option than adding more server nodes. Open-source solutions may someday provide additional choices in this area.

HMM searches are powerful tools, and the proper system can permit the application of those tools to large datasets. Time spent on a few careful calculations of future needs can save a greater amount of time in the long run as those needs tend to expand and become more complex over time.

7.8 QUESTIONS

1. An overworked graduate student has 10,000 protein sequences to analyze with the TIGRfam database. She wants to identify the number of potential tyrosine kinases in this set, but has no access to an accelerator and her lab cluster is busy running other jobs. Left with an aging server with which to run HMMer, how should she go about her analysis in a timely fashion?

2. Download the optimized HMMer code from scalableinformatics.com and install this on any Linux system to which you have access. Run this code with a dataset provided by your instructor against the Pfam database, and compare the runtime to that of the standard HMMer distribution for that dataset.

3. Download the trial version of the optimized HMMPfam code from sp-soft.com and install this on any Linux system to which you have access. Run this code with a dataset provided by your instructor against the Pfam database, and compare the runtime to that of the standard HMMer distribution for that dataset.

4. A professor wants to analyze 10,000 ESTs against the Pfam database to provide preliminary data for an upcoming grant proposal. Discuss the advantages and disadvantages of using ESTwise versus translating the sequences into proteins and running them with hmmPfam.

5. Run a set of 100 ESTs provided by your instructor against the CDD database at NCBI using RPS-BLAST. Then run the same data on the GeneWise server at http://www.ebi.ac. uk/Wise2/. Compare and contrast the output in terms of what was hit, what was missed, scores and E-values.

References

[1] G. Chukkapalli, C. Guda and S. Subramaniam, "SledgeHMMER, a Web Server for Batch Searching the Pfam Database," *Nucleic Acids Res.* **32**(web server issue), W542–W544 (2004).

[2] J. Walters, B. Qudah, and V. Chaudhary, "Accelerating the hmmer Sequence Analysis Suite Using Conventional Processors," (2006), *Proceedings of the 20th International Conference on Advanced Information Networking and Applications Volume 1 (AINA'06)*, pp. 289–294, Washington, D.C., IEEE Computer Society.

[3] B. Wun, J. Buhler, and P. Crowley, "Exploiting Coarse Grained Parallelism to Accelerate Protein Motif Finding with a Network Processor," 2005, *PACT '05: Proceedings of the 2005 International Conference on Parallel Architectures and Compilation Techniques.*

[4] J. Landman, J. Ray, and J. Walters, "Accelerating hmmer Searches on Opteron Processors with Minimally Invasive Recoding," *Proceedings of the 20th International Conference on Advanced Information Networking and Applications Volume 2* (AINA'06), pp. 628–636, Washington, D.C., IEEE Computer Society.

[5] E. Lindahl, "Altivec Accelerated HMM Algorithms," 2005. http://lindahl.sbc.su.se/.

[6] D. Horn, M. Houston, and P. Hanrahan, "ClawHMMER: A Streaming HMMer-Search Implementation," *Supercomputing, 2005, Proceedings of the ACM/IEEE SC 2005 Conference*, 12–18 Nov. 2005, pp. 11–11.

[7] E. Birney, M. Clamp, and R. Durbin, "GeneWise and Genomewise," *Genome Res.* **14**(5), 988–995 (2004).

[8] V. Curwen, E. Eyras, T. Andrews, L. Clarke, E. Mongin, S. Searle, and M. Clamp, "The Ensembl Automatic Gene Annotation System," *Genome Res.* **14**(5), 942–950 (2004).

[9] Progeniq is the company that makes the BioBoost accelerator system. See their product line at www.progeniq.com.

[10] TimeLogic makes the DeCypher line of accelerators that include accelerated HMM searches. See www.timelogic.com.

Chapter 8

Other Uses of HMMs in Bioinformatics

8.1 INTRODUCTION

We have concentrated on Profile-HMM searches in general, and HMMer in particular, because this is where the biggest need for instruction exists. Hidden Markov Models are not stagnant concepts by any means, and work is constantly being done in the field. As we have seen in chapter 7, a great deal of work is directed towards speeding up these searches in a wide variety of methods. An equal amount of work is being done to expand capabilities, increase sensitivity, and improve scoring.

Hidden Markov Models may be used to predict specific features about a sequence, the cellular location, or to predict post-translational modifications. They may help in generating improved alignments and in making fold structure predictions.

HMMs are a versatile means for elucidating information about the overall biology of a cell or an entire system, and the possibilities have not yet begun to be tapped out. We have so far been interested in Profile-HMMs that are first order models, but there may be some benefit to higher order models in various bioinformatics studies. Higher order HMMs may be able to capture longer-range effects having to do with protein structural characteristics, although so far experiments have not been successful in this area.

We will also examine the MEME/MAST/Meta-MEME suite of programs that are used for building motifs, searching for additional proteins or nucleotides, and building Hidden Markov Models from those motifs. This system looks for ungapped motifs in a set of training data based on parameters selected by the user.

8.2   METHODS COMPARING HMMS TO OTHER HMMS

All of the algorithms that we have examined so far have had one feature in common: they compare a Profile-HMM and a sequence. There may be notable differences in how they go about it, with various sorts of preprocessing steps and levels of complexity in the calculations, but eventually they all get down to the alignment of the sequence to the model.

Methods for the comparison of Profiles to other profiles such as COMPASS and prof_sim have shown that some of the models in Pfam have distant relationships with other family or domain models. This is discovered to be the result of structural relationships.

8.2.1   PRC

The profile comparer, PRC, by Martin Madera, is designed to compare profiles of several different types to other profiles. This has been shown to be extremely sensitive at detecting remote relationships compared to profile-sequence comparisons (which are themselves much more sensitive than sequence-sequence comparisons).

PRC compares all probabilities, both transition and emission, in both Hidden Markov Models that are being compared. PRC is format agnostic, and can read HMMer, SAM, PSI-BLAST and FASTA files. Madera recommends using the SAM package to build the Profile-HMMs.

PRC can run in pairwise mode, where an HMM is compared to another, or in a batch library mode, where a model is compared to an entire database of models. This is particularly useful when you have generated a new model and want to see if it is similar to anything in the Pfam (or other) HMM database.

PRC can use the viterbi or the forward algorithm. The forward algorithm analyzes all possible alignments, where viterbi is trying to find the single best alignment. The forward is generally considered to be superior to Viterbi, although, as you might suspect, it is considerably slower.

PRC is capable of running in any combination of global and local modes. Using the command

$$\text{prc -mode local-global test1.hmm test2.hmm}$$

will produce an alignment that is local to test1.hmm and global to test2.hmm

PRC can utilize cutoff thresholds with the use of the -Emax option. If the option -Emax 0.01 is used, then any alignments with evalues higher than this will be discarded. The default value for this setting is 10.

PRC uses the Waterman-Eggert method to find all alignments between a pair of models. This can result in some long runtimes, as it can always try to find more alignments. To prevent endless loops, put in a stop value (which is basically a score threshold) using the -stop option, or limit the total number of hits with the -hits option.

8.2.2 HHsearch

HHsearch by Soding compares HMMs to other HMMs, rather than using profiles as do COMPASS and prof_sim. As a result, HHsearch is reported to find significantly more homologs at a given rate of false positives, and dramatically more than PSI-BLAST and HMMer. Still, HHsearch is faster than most other methods of this type.

HHsearch and the associated website called HHpred are designed to be fast, simple to use and interpret the output, and flexible. As an example of the flexibility, the user of the HHpred database can paste in his own query alignment, select either global or local alignment mode, and change parameters on the fly to see how the alignment is altered.

The method of HHsearch has been extended in several ways. HHrep is designed to find repeat regions (de novo), HHOMP is designed specifically for the identification of OMP proteins (Outer Membrane Proteins) and HHalign provides alignments based on the input model.

8.2.3 SCOOP

SCOOP is an elegant method of comparing two profiles. Rather than compare them directly, the two Profiles are compared to a protein database. Different targets may be used, but they should be rather large—over 1,000,000 sequences for best results. The output files of the two runs are compared to one another, even those hits of low quality that would normally be discarded. If the two Profiles have more hits in common than would be expected by chance, then they are considered to be related. This is an interesting idea that performs rather well, and in comparisons between SCOOP, PRC and HHsearch, the interesting thing is that the methods appear to be nicely complementary. This means that a metasearch tool could be constructed that would combine the best parts of all of them, for much better results than any of them could achieve alone.

8.2.4 More Profile Comparison Systems

Other Profile-Profile systems that are designed for building the most accurate MSAs in the same vein as HHalign are SATCHMO by Sjolander and Edgar and PCMA by Pei et al.

The COACH tool (Comparison of Alignments by Constructing Hidden Markov Models) from Sjolander and Edgar is a different approach to Profile-Profile alignment. The COACH algorithm takes one alignment, builds a Profile-HMM out of it and then aligns the other to the HMM. The authors of the COACH paper have compared the performance of this algorithm to those of prof_sim and COMPASS, and found that their method produces better results, on average, than either of the other two systems.

8.3 SUBCELLULAR LOCALIZATION PREDICTION

The pTARGET system by Guba and Subramaniam is designed to take a protein sequence and predict the location within the call that the protein might be found. The authors found a set of Pfam models that are strongly associated with a particular location within the cell. For example, the homeobox domain is nucleus-specific. If the sequence hits the homeobox domain model in Pfam, then it is likely that the protein is found in the nucleus.

Next, the amino acid composition is considered. The composition of protein sequences in the various locations has been analyzed, and the composition contributes to the overall score. The weighting of the Pfam score and the composition score depend on the signals from each. For example, a clear signal from the Pfam database will take precedence over several signals from the amino acid composition.

Other subcellular location prediction programs that rely in part on HMMs include mitopred, mitpred and signalP. The SignalP algorithm is currently the most popular method for predicting secretory proteins, and is based on an interesting combination of HMMs and artificial neural networks.

Transmembrane proteins are extremely important in a number of biological systems such as cellular signalling, transport and ion channels. Anders Krogh and Erik Sonnhammer created the TMHMM program to predict these types of proteins. The models were trained on 160 membrane proteins, and a set of 645 non-membrane proteins was used as a negative test set to determine the discrimination power of the algorithm. The TMHMM website at http://www.cbs.dtu.dk/services/TMHMM-2.0/ produces output in graphical or text format that shows predictions for regions as either inside, outside or transmembrane helix.

Later HMM-based transmembrane prediction programs include HMMTOP, S-TMHMM, PRO-TMHMM and PRODIV-TMHMM.

PRED-TMBB is an HMM-based model that can predict the transmembrane beta barrel strands of the outer membrane proteins from gram-negative proteins. PRED-TMBB can discriminate between these beta-barrel proteins and water soluble proteins in large screens.

8.4 POST-TRANSLATIONAL MODIFICATION PREDICTION

Proteins can be modified in several ways after translation from mRNA. These modifications may be seen as a later stage of protein biosynthesis, and the range of function of the protein is extended considerably by the addition of functional groups such as acetate, phosphate, sulfate and various lipids and carbohydrates as well as dozens of other changes that can be made to the protein. Some of these modification sites can be predicted with reasonable confidence, and naturally Hidden Markov Models are frequently involved with these predictions.

For proteins that go through the secretory pathway, tyrosine sulfation is a key post-translational modification where a sulfate group is added to a tyrosine. The extracellular portions of a membrane protein that pass through the Golgi apparatus are also candidates for sulfation.

A program called 'sulfinator' utilizes 4 HMMs that were trained on a MSA of 68 windows around known sulfation sites. Each window was 22 residues in length. The models were then tested on an additional 121 known sulfation sites and 183 negative sites.

The sulfinator is available online at the Expasy website at http://www.expasy.ch/tools/sulfinator/.

Other post-translational modification prediction algorithms that are based on Hidden Markov Models include sulfosite, Glycosee, and KinasePhos, which together make up the dbPTM system.

8.5 BINDING SITE PREDICTIONS

Andreas Henschel and his colleagues at the Biotech Center in Dresden, Germany, have taken the protein-protein interactions and protein-ligand interaction information from the Protein DataBank (PDB). They have constructed these into a set of HMMs that describe these sites—740 for the protein-protein binding sites and 3,000 for the protein-ligand binding sites.

The protein-protein interactions are taken from SCOPPI, the Structural Classification of Protein-Protein Interfaces. SCOPPI is built by the same team in Dresden, by making structural alignments of SCOP families. The models were validated in two ways: First, the models were cross-validated with structure data sets. Secondly, the dataset was tested against literature-curated interactions. For this test, the authors turned to netPro, an expertly curated database with information on about 15,000 protein-protein interactions.

The protein-ligand interface models were built by scanning the PDB for crystal faces that bind to small molecules and peptides, and building models for these. To accomplish this, the authors found the most commonly occurring ligands that were co-crystallized with the associated protein structures. The models were created from the residues that surrounded the ligand, including any possible cofactors.

The models representing both protein-protein interaction sites and protein-ligand interaction sites are freely available from the authors if you are an academic.

8.6   GENE FINDING PROGRAMS

The hidden Markov Models that we have looked at throughout the course of this text have used amino acids in each state. Other uses for HMMs in bioinformatics involve the modeling of a series of features rather than a series of letters. These features might include promoter regions, exons, introns, Poly-A tails and TATA boxes, among others. This is the ability to model grammar, instead of individual letters. Certain rules for these features can improve the modeling, such as the minimum length of Exons, the end of the gene cannot be an intron, and so forth.

Hidden Markov Models have been used in gene finding algorithms since 1994. Gene finders use a variation on the standard Hidden Markov Model methodology called General Hidden Markov Models (GHMMs). The more familiar pair HMMs may be combined with this method to form a Generalized Pair Hidden Markov Model (GPHMM).

As a sequence is run through the genefinder, different parts or regions of the sequence are assigned to features with associated probabilities.

A Danish group has developed a gene finding system called Evogene that models both genome structure and evolution. This has been named an Evolutionary Hidden Markov Model, or EHMM, because it consists of an HMM and a group of evolutionary models that use a phylogenetic tree as a foundation. Evogene is designed to handle large numbers of aligned genomes, as long as they are linked with a phylogenetic tree to model the evolutionary correlations.

Other HMM-based genefinders include GenomeScan, Twinscan, Snap, Genezilla, Augustus and FgenesH. Various studies have been made about the overall accuracy of these programs. The results about which is 'best' depends upon the metric that is chosen. One might perform better at the exon level, worse at the nucleotide level, and in the middle of the pack in terms of whole genes.

As the quality of Profile-HMMs depends largely on the training data, so does the power of a genefinder depend greatly upon the type of organisms that are used in the model training. Therefore, a genefinder that has been trained on mammalian genomes will not perform as well on average when used on plant data.

8.7 MEME, MAST AND METAMEME

While MEME (which stands for Multiple Expectation Maximization for Motif Elicitation) is not strictly an HMM implementation, the associated MetaMEME program is considered to be in this category. MEME is a method for discovering motifs in a set of related sequences. One of the interesting things about MEME is that it is designed to work with either protein or nucleotide data (although protein is the default).

MEME finds profiles with no gaps. Sequence patterns that contain gaps are split by MEME into separate motifs. The maximum number of motifs is set by the user—the width, number of hits, and other details about the motifs are found automatically.

MEME may be used with the following syntax:

> meme lea.faa -minw 5 -maxw 20 -nmotifs 5 -mod anr
>
> > lea_motifs.html

The format is amino acid, the name of the sequence file is lea.faa. Setting the minimum and maximum widths of the motifs that are being searched will save time, because regions greater and smaller than these values will not be examined. Much greater speed could be achieved by using the -w option, which forces the motifs to have a specific width. Limiting the total number of motifs to 5 will also speed execution of the program.

On the other hand, setting the -mod to anr means that there may be any number of repeats. This is necessary if you are searching for repeat domains, but will slow the program runtime considerably.

The default output format is html; setting the -txt switch will force output into plain ASCII text format. Using the -dna option will enable the program to utilize nucleotide sequences, as in the following example:

> meme test.fna -dna -revcomp -txt -mod oops > test_motifs.txt

The -revcomp option may be used with nucleotide data to search both forward and reverse complement strands. The 'oops' model restricts the program to assume that each sequence contains only one occurrence of the motif.

The following command,

> meme TKR.faa -mod zoops -nmotifs 10 -maxw 30 -evt .001
>
> > TKR_m.txt

will tell meme to look through the TKR.faa file for up to 10 motifs, with zero or one per sequence, of a width no greater than 30, and an E-value no greater than .001.

meme will output:

- The version and release date of MEME.

- The reference to cite in your publications

- Information about the training set—that is, the sequences that you submitted.

- The command line that you used to run MEME.

- Summary information about the motif.

- A simplified scoring matrix

- A diagram showing the degree of conservation of residues at each position

- A consensus sequence, showing the most conserved residue at each position

- Diagrams of the motif as it occurs within each sequence of the training set.

- The motif in BLOCKS format.

- The Profile or PSSM for use by MAST

- The motif described as a Position Specific Probability Matrix, or PSPM.

- A summary of all the discovered motifs showing the tiling over each sequence in the training set.

8.7.1 MAST (Motif Alignment and Search Tool)

MAST will search a protein or nucleotide database with an input file containing one or more motifs, generating output as seen in figure 8.1. The motif input file may be generated by MEME or any other program that produces the same format.

The command

        mast lea_motifs.html -d uniref90

will search the nonredundant uniprot database with the the lea motifs that we generated earlier. To limit the search to the first two models in the lea_motifs.html file, we can add the -c option. To set a significance threshold of evalue 0.01, we may use the -ev option as in the following command:

        mast lea_motifs.html -d uniref90 -c 2 -ev .01

Links	Name	Expect	Motifs
§A?	At1g01140.1_4-2-4_SnRK3.12	5.5e-75	
§A?	At1g01140.2_SnRK3.12	5.6e-75	
§A?	At1g01140.3_SnRK3.12	5.7e-75	
§A?	At1g06390.2_ASK-iota	1.6e-65	
§A?	At1g06390.1_4-5-4_ASK-iota	1.6e-65	
§A?	At1g01560.1_4-5-1_MPK11	6.5e-53	
§A?	At1g04440.1_3-1-1-1	2.9e-50	
§A?	At1g03930.1_3-1-1-1_ADK1	2.5e-49	
§A?	At1g03920.1_4-2-6	1.2e-46	
§A?	At1g01540.1_1-6-3	2.5e-34	
§A?	At1g01540.2_Putative	4.8e-34	
§A?	At1g03740.1_4-5-2	2.6e-31	

FIGURE 8.1: MAST Output

MAST has a couple of particularly interesting features. In the following command,

mast lea_motifs.html -d est_others -dna -comp

the -dna option specifies that the DNA sequences are converted into protein in order to be searched by the protein motifs. The -comp option adjusts the E-values for sequence composition, something like the Null2 scoring correction feature in HMMer.

8.7.2 Meta-MEME

The motifs that were discovered earlier by MEME may now be tied together into a Hidden Markov Model with a program called Meta-MEME. The inputs are the MEME motif models, which are gapless PSSMs. The sequence file from which these models were derived is the other input.

Meta-MEME takes the single-motif models from the MEME HTML file and builds them into a multi-motif Hidden Markov Model. This adds the transition probabilities to allow for insertions, deletions and repeated domains.

Since the MEME motifs are gapless, the regions between the motifs are modeled imprecisely by design. This means that, since we are not worried about the regions between the motifs, we may train Meta-MEME models with fewer sequences than would be required by HMMer or SAM.

Our newly created MetaMEME model can now be used to search a sequence database for homologs. As with HMMer, either the Viterbi algorithm or the forward algorithm may be used. The Meta-MEME bitscore is a log-odds score. This score is formed by the ratio of the score of the sequence and the foreground model versus the score of the background model. The bit score is the base 2 logarithm of this ratio. The foreground model is an HMM, and the background model is representative of a random sequence. Typically, a bit score of 0 would indicate that there is a 50% chance that the sequence belongs to that family.

In practice, the statistical significance of a bit score depends on the number of family memebers that you would expect to find in the database. If your

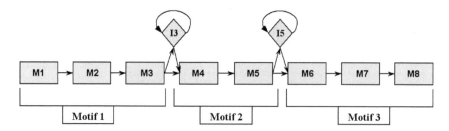

FIGURE 8.2:   Three Tiny Example Motifs Built into a Meta-MEME HMM

database contains 1,000,000 sequences, and the family has 100 members in the database, then the odds score threshold would be $1,000,000/100 = 10,000$ or greater would likely be a member of that family. The base 2 log score would then be 13.3.

8.7.3   Meta-MEME Programs

Meta-MEME consists of four basic programs and several auxiliary programs. The overall process is shown in figure 8.3.

8.7.3.1   MHMM

MHMM builds the model from the MEME motifs. Due to the architecture, the models have a few different options than other systems such as SAM or HMMer. The models may be linear, with each motif following the other like a set of beads on a necklace, with each motif connected with inserts that represent the regions between the motifs, as shown in figure 8.2. A connected model differs from the linear in that it has a more generalized topology. The output of each motif is connected to the input of every other motif. This format allows for repeat motifs, order shuffling and deleted motifs.

The order of the motifs in a Meta-MEME HMM and the transition probabilities between the motifs are derived from the MEME file and the highest-scoring sequence for the default linear models. For models of the 'completely connected' type, the order is taken from all of the training sequences. If you would like to override the order, the –order option may be used to specify it:

mhmm –type complete lea_motifs.meme > lea_motifs.hmm

8.7.3.2   MHMMS

The Hidden Markov Model generated by the mhmm command is run against a FASTA sequence database using the mhhms executable. This is roughly the equivalent of the hmmsearch command in the HMMer world. No formatting

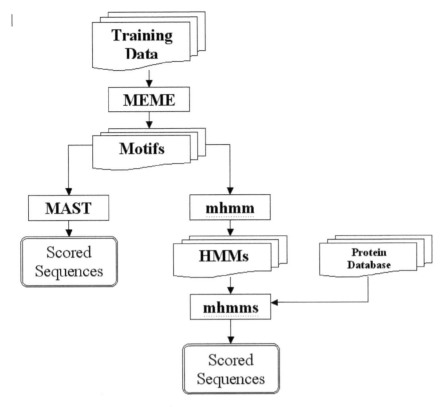

FIGURE 8.3:   The MEME/MAST/Meta-MEME Process

of the database is required. Each sequence in the database is assigned an E-value, and the sequences scoring below the specified threshold (that is, better than the threshold), are reported:

mhmms lea_motifs.hmm uniprot.fa > lea_hits.out

8.7.3.3 mhmmscan

While mhmms is the standard for searching most motifs against most databases, such as a normal nucleotide motif against an EST database, for example, there are limitations to the program. Using mhmmscan is like using mhmms except for two key differences. First, the mhmmscan program can take long sequences, such as contigs or chromosomes. Second, it can match a model to the sequence many times, which is similar to turning of the J state in HMMer.

These two additions are important to genome annotation. If the model is of a type that is likely to be found frequently like a repeat region, or if it is of a common type such as a phosphorylation prediction model, then mhmmscan will be the program of choice. On the other hand, if you only want a single hit per sequence, then you want to use mhmms:

Mhmmscan mydna.hmm Arabidopsis.fnt > mydna_hits.out

8.7.3.4 Tom-tom

This is actually a Profile-Profile comparison tool in the same vein as PRC, HHsearch or SCOOP. The difference is that tom-tom is designed for MEME style motifs. With the tom-tom program, the query is a motif that is run against a database of target motifs.

For a given pair of motifs, a minimum number of positions is required for overlap. For the overlapping positions, scores are compared against the background amino acid frequencies from the MEME input files. Scores are generated for every column that overlaps, and these scores are summed. The overall score is converted to an E-value, and the lowest E-value over all offsets is reported back to the user.

8.8 CLUSTALW

We will now undertake a brief introduction to MSA with ClustalW. While other programs are now available, ClustalW is still the most popular for this type of work.

ClustalW is a commonly used general-purpose MSA program for DNA or proteins. As an intermediate step in the production of HMMs, it is commonly used for protein alignments. It lines up the selected sequences so that the identities, similarities and differences are easily viewed. Cladograms or Phylograms are then used to show evolutionary relationships.

Consensus symbols rather than consensus sequences are used to represent the degree of conservation in each column. A '*' (asterisk) indicates that the residues are identical (that is, they are completely conserved) in that column. A ':' (colon) indicates that one of the following strongly conserved substitutions is found in that column:

FYW

HY

MILF

MILV

NEQK

NHQK

NDEQ

STA

QHRK

A '.' (period) indicates that one of the following weakly conserved substitutions is found in that column:

ATV

CSA

FVLIM

HFY

NDEQHK

NEQHRK

SAG

SGND

SNDEQK

STNK

STPA

Several websites have been established to enable Internet access to Clustal-W. These are easy to use, and can have some additional features that are not found in the command line version, but they are usually limited in the number and length of sequences that may be submitted. Columns are sometimes color coded to indicate conservation. A red column indicates conservation between the small and hydrophilic residues (AVFPMILW and Y).

Blue columns indicate acidic residues (D and E). Magenta columns are indicative of basic residues (R, H and K). Basic residues that also have hydroxyl and amine groups (STYHCNG and Q) are colored in green. All other groups are colored in grey.

The following table provides a list of file types that are understood by ClustalW and the first characters in those files that identify the type:

Format	First Character or Characters
Fasta	>
NBRF	>P1;
EMBL/Swiss	ID
GDE protein	%
CLUSTAL	CLUSTAL
GCG/MSF	PILEUP or !!AA_ or MSF
GCG9/RSF	!!RICH_SEQUENCE

ClustalW uses these characters to identify the sequence format, and uses character frequencies to identify whether the sequences are amino acid or nucleotide. If more than 85% of the characters are A,C,T,G, U and N, then DNA or RNA is assumed. If less than 85% of the sequence is made up of these six characters, then it is assumed to be an amino acid sequence.

ClustalW can also output to a number of different alignment formats. For the production of HMMs, the default clustal alignment (with extension .aln)

```
(
(
(
gi—122615—sp—P02023—HBB_HUMAN:0.08397,
gi—122614—sp—P02062—HBB_HORSE:0.08042)
:0.23814,
(
gi—122412—sp—P01922—HBA_HUMAN:0.06074,
gi—122411—sp—P01958—HBA_HORSE:0.05898)
:0.20431)
:0.07775,
(
gi—127687—sp—P02185—MYG_PHYCA:0.37678,
gi—126238—sp—P02240—LGB2_LUPLU:0.47943)
:0.01624,
gi—121233—sp—P02208—GLB5_PETMA:0.37780);
```

FIGURE 8.4:   Example of the New Hampshire Format

is sufficient and easy to read. Jalview is a powerful and flexible program for viewing and editing these alignments. One problem with Jalview is that it does not allow you to save and print the image. This is easily rectified, however, with the use of a screen grab program. There may be times when we wish to use another program to view and manipulate the alignments, hence the multiple output options. In addition to the default blocked alignment, ClustalW is capable of outputting to NBRF/PIR format, which uses dashes for gaps; MSF format, which is the standard GCG package format; PHYLIP; and GDE (Genetic Data Environment). You also have the option of having the sequences in the same output order as they were entered, or writing them out in the order that they were aligned.

The Clustalw phylogenetic tree file has an the extension of '.dnd,' which follows the New Hampshire format of nested parentheses (see figure 8.4). ClustalW does not provide any method for viewing these trees graphically. There are many other programs that enable the graphical viewing of New Hampshire format trees, such as treeview (see figure 8.5), ATV (A Tree Viewer) and hypertree.

When you view these trees, you will encounter the terms phylogram and cladogram. A phylogram is a branching tree that is assumed to be an estimate of a phylogeny. The amount of evolutionary change is inferred from the branch lengths.

In a cladogram, the branches are of equal length. A cladogram shows common ancestry, but gives no indication of the amount of evolutionary time that separates the taxa.

FIGURE 8.5: Treeview Output

8.8.1 Advanced ClustalW Options

- Alignment:
  You may choose to run a full alignment or use a stringent algorithm for generating the tree guide or a fast algorithm.

- Output:
  Here you decide which output format you want your MSA in. The options are ALN, GCG, PHYLIP, GDE and PIR. Typically, we will want to stick with ALN for the building of models, although at times we may want to bring the alignments into PHYLIP to better determine where to split the data, if necessary.

- Outorder:
  Decide which order the sequences should be printed in the alignment.

- Fast Pairwise Alignment Options:

  (a) KTUP - This option allows you to choose which 'word-length' to use when calculating fast pairwise alignments. (Note: make sure you have chosen 'fast' in the alignment.)

  (b) Window - Use this option to set the window length when calculating fast pairwise alignments. (Note: make sure you have chosen 'fast' in the alignment.)

  (c) Score - This option allows you to decide which score to take into account when calculating a fast pairwise alignment. (Note: make sure you have chosen 'fast' in the alignment.)

  (d) Topdiag - Select here how many top diagonals should be integrated when calculating a fast pairwise alignment. (Note: make sure you have chosen 'fast' in the alignment.)

  (e) Paigap - Select here to set the gap penalty when generating fast pairwise alignments.

- Matrix:
  This option allows you to choose which matrix series to use when generating the MSA. The program goes through the chosen matrix series, spanning the full range of amino acid distances.

(a) Blosum (Henikoff) - These matrices appear to be the best available for carrying out database similarity searches. The matrices used are: Blosum80, 62, 40 and 30.

(b) Pam ((Dayhoff) - These have been extremely widely used since the late '70s. Clustal uses the PAM 120, 160, 250 and 350 matrices.

(c) Gonnet - These matrices were derived using almost the same procedure as the Dayhoff one (above) but are much more up to date and are based on a far larger data set. They appear to be more sensitive than the Dayhoff series. We use the GONNET 40, 80, 120, 160, 250 and 350 matrices.

- Gaps:

(a) Gapopen - You can set here the penalty for opening a gap. The default value is 10.

(b) Endgap - You can set here the penalty for closing a gap.

(c) Gapext - You can set here the penalty for extending a gap. The default value is 0.05.

(d) Gapdist - You can set here the gap separation penalty. The default value is 8.

8.8.1.1   Alternatives to ClustalW

Several alternatives exist for those who wish to build MSAs. Two goals are paramount in this endeavor, the accuracy of the resulting alignment and the amount of time that the alignment requires.

Kuo-Bin Li created a distributed and parallelized version of ClustalW that can speed up large alignment jobs across a cluster of computers using MPI. All three of the steps have been optimized—the initial pairwise alignment, the building of the guide tree and the progressive alignment.

TCoffee is designed to improve the quality of the alignments by including information from multiple sources, such as both global and local alignments. Although TCoffee is considered to be very accurate, the runtime is proportional to the cube of the number of sequences. This makes it very time consuming to analyze larger families in most servers.

Dozens of other MSA programs have been developed—over 50 in the past decade. Some, such as MAFFT and PRALINE, start by searching a database to extract homologous sequences. These sequences are used to help produce the alignments, and then are removed for the final presentation.

Expresso (which has replaced 3Dcoffee) is based on TCoffee, and combines structural alignment with the threading program Fugue with the pairwise sequence alignment to produce a more accurate overall alignment. For distantly related sequences, more than one structure should be utilized, with a consequent increase in runtime.

8.9   SUMMARY

In chapter 8, we have seen a number of variations to the standard Profile-HMM sequence analysis methods. Profile-Profile methods can be extremely sensitive methods for remote homology detection, and several of these methods are now available.

The prediction of subcellular localization is useful in understanding the overall picture of the function of a protein. Mitopred, mitpred, Ptarget and signalP are predictors that are at least partially based upon HMM searches.

Transmembrane proteins are important candidates for drug targets. Hidden Markov Model predictors for these types of proteins model the alternating hydrophilic and hydrophobic regions rather than the individual amino acids.

Protein binding site prediction may also accomplished with a set of specific HMMs. This set of HMMs that describe these sites—hundreds for the protein protein binding sites and several thousands for the protein-ligand binding sites.

Post-translational modification is the chemical alteration of a protein. HMMs are excellent tools for the discovery of PTMs of different sorts in different types of organisms.

Gene finders are roughly split into de-novo and homology-based tools. Many of the modern programs combine the two techniques. Of the two types, HMMs are utilized in the de-novo methods to model the various features of the genes.

Meta-MEME is a method for building and searching with motif-based HMMs. Regions between blocks are less precisely modeled than in other systems, which makes it easier to model families with fewer members.

Other uses for HMMs in bioinformatics include:

- Protein fold recognition
  Karplus and many others have proven the utility of HMMs in predicting secondary structure. Structural HMMs have been shown by Bouchaffra and Tan to be superior to Support Vector Machine methods in the prediction of protein folds.

- Detection of CpG islands
  In mammalian DNA, the dinucleotide CG is relatively rare. The reason for this is that the C in this type of pairing will become methylated, that is, one of the hydrogens will be replaced by a $-CH_3$(Methyl) group. This creates a high probability that the C will mutate to a T, and the CG (or CpG) dinucleotide is seldom found. This methylation process is suppressed upstream of a gene, and so the identification of a CpG island, where the numbers of the CG dinucleotide are relatively high, is an indicator of a gene.

- Signaling region identification
  Signal peptides have a characteristic structure that consists of 1-5 mostly positively charged amino acids, followed by 7-15 hydrophobic residues, and finally 3-7 polar amino acids. The Hidden Markov Models are used to identify these peptides, identify a cleavage site and discriminate between secretory and non-secretory proteins.

8.10　QUESTIONS

1. What is the relationship between Genscan, GenomeScan and TwinScan?

2. Submit 20 sequences that have been supplied by your instructor to the KinasePhos website. How many phosphorylation sites are predicted in this dataset?

3. Given a set of protein sequences, choose three different websites for prediction of transmembrane proteins, and report back on the results—how did they compare, which reported the most TM regions, and whether there seemed to be consensus on the predictions.

4. Submit a set of DNA sequence training data provided by your instructor to the MEME website. Choose any number of repetitions, minimum width of 6 and maximum width of 50, with a maximum number of 3 motifs found.

5. Take the output from problem 3 and submit it to the MAST search website. Choose PDB as your target database, and leave all the other options at their defaults.

6. Build the motifs from problem 3 into a Hidden Markov Model at the Meta-MEME website. Choose the NCBI non-redundant database, and ask for the following output:

   A. A motif-based HMM.

   B. A list of homologous sequences from the NCBI-NR database

References

[1] N. Dickens and C. Ponting, "THoR: a Tool for Domain Discovery and Curation of Multiple Alignments," *Genome Biol.* **4**(8), R52 (2003).

[2] A. Henschel, C. Winter, W. Kim, and M. Schroeder, "Using Structural Motif Descriptors for Sequence-Based Binding Site Prediction," *BMC Bioinformatics.* **8**(Suppl 4), S5 (2007).

[3] P. Srivastava, D. Desai, S. Nandi, and A. Lynn, "HMM-ModE— Improved Classification Using Profile Hidden Markov Models by Optimising the Discrimination Threshold and Modifying Emission Probabilities with Negative Training Sequences," *BMC Bioinformatics* **8**,104 (2007).

[4] A. Bateman and R. Finn, "SCOOP: a Simple Method for Identification of Novel Protein Superfamily Relationships," *Bioinformatics* **23**(7), 809–814 (2007).

[5] D. Haussler, "Computational Genefinding," *Trends Biochem. Sci.* 12–15 (1998).

[6] C. Burge and S. Karlin, "Finding the Genes in Genomic DNA," *Curr. Opin. Struct. Biol.*, **8**, 346–354 (1998).

[7] M. Burset and R. Guigo, "Evaluation of Gene Structure Prediction Programs," *Genomics* **34**, 353–367 (1996).

[8] S. Rogic, A. Mackworth, and F. Ouellette, "Evaluation of Gene-Finding Programs on Mammalian Sequences," *Genome Res.* **11**, 817–832 (2001).

[9] S. Salzberg, A. Delcher, S. Kasif, and O. White, "Microbial Gene Identification Using Interpolated Markov Models," *Nucleic Acids Res.* **26**, 544–548 (1998).

[10] A. Lukashin and M. Borodovsky, "GeneMark.hmm: New Solutions for Gene Finding," *Nucleic Acids Res.* **26**, 1107–1115 (1998).

[11] J. Henderson, S. Salzberg, and K. Fasman, "Finding Genes in Human DNA with a Hidden Markov Model," *J. Comput. Biol.* **4**, 127–141 (1997).

[12] L. Pachter, M. Alexandersson, and S. Cawley, "Applications of Generalized Pair Hidden Markov Models to Alignment and Gene Finding Problems," *J. Comput. Biol.* **9**, 389–399 (2002).

[13] A. Krogh, I. Mian, and D. Haussler, "A Hidden Markov Model that Finds Genes in E. coli DNA," *Nucleic Acids Res.* **22**, 4768–4778 (1994).

[14] D. Kulp, D. Haussler, M. Reese, and F. Eeckman, "A Generalized Hidden Markov Model for the Recognition of Human Genes in DNA," *Proceedings of the Fourth International Conference on Intelligent Systems for Molecular Biology*, AAAI Press, Menlo Park, CA, pp. 134–142 (1996).

[15] A. Krogh, "Two Methods for Improving Performace of a HMM and Their Application for Gene Finding," *The Fifth International Conference on*

Intelligent Systems for Molecular Biology, AAAI Press, Menlo Park, CA, pp. 179–186 (1997).

[16] M. Borodovsky, J. McIninch, E. Koonin, K. Rudd, C. Mdigue, and A. Danchin, "Detection of New Genes in a Bacterial Genome Using Markov Models for Three Gene Classes," *Nucleic Acids Res.* **23**, 3554–3562 (1995).

[17] Z. Wang, Y. Chen, and Y. Li, "A Brief Review of Computational Gene Prediction Methods," *Genome Proteomics Bioinformatics* **2**, 216–221 (2004).

[18] W. Majoros, M. Pertea, A. Delcher and S. Salzberg, "Efficient Decoding Algorithms for Generalized Hidden Markov Model Gene Finders," *BMC Bioinformatics* **6**, 16 (2005).

[19] M. Stanke and S. Waack, "Gene Prediction with a Hidden Markov Model and a New Intron Submodel," *Bioinformatics* **19**, ii215–ii225 (2003).

[20] W. Majoros, M. Pertea, and S. Salzberg, "TigrScan and GlimmerHMM: Two Open Source ab Initio Eukaryotic Gene-Finders," *Bioinformatics* **20**, 2878–2879 (2004).

[21] I. Korf, "Gene Finding in Novel Genomes," *BMC Bioinformatics* **5**, 59 (2004).

[22] K. Knapp, K. Chen, and Y. Phoebe, "An Evaluation of Contemporary Hidden Markov Model Genefinders with a Predicted Exon Taxonomy," *Nucleic Acids Res.* **35**(1), 317–324 (2007).

[23] W. Majoros, M. Pertea, C. Antonescu, and S. Salzberg, "GlimmerM, Exonomy and Unveil: Three ab Initio Eukaryotic Genefinders," *Nucleic Acids Res.* **31**(13), 3601–3604 (2003).

[24] T. Bailey and C. Elkan, "Fitting a Mixture Model by Expectation Maximization to Discover Motifs in Biopolymers," *Proceedings of the Second International Conference on Intelligent Systems for Molecular Biology* AAAI Press, Menlo Park, CA, pp. 28–36 (1994).

[25] T. Bailey and M. Gribskov, "Combining Evidence Using p-Values: Application to Sequence Homology Searches," *Bioinformatics* **14**, 48–54 (1998).

Index